THE UNIVERSITY OF WESTERN ONTARIO SERIES IN PHILOSOPHY OF SCIENCE

A SERIES OF BOOKS IN PHILOSOPHY OF SCIENCE, METHODOLOGY, EPISTEMOLOGY, LOGIC, HISTORY OF SCIENCE, AND RELATED FIELDS

VOLUME 25

SCIENTIFIC RATIONALITY: THE SOCIOLOGICAL TURN

SCIENTIFIC RATIONALITY: THE SOCIOLOGICAL TURN

JAMES ROBERT BROWN
Department of Philosophy, University of Toronto, Canada

D. REIDEL PUBLISHING COMPANY
A MEMBER OF THE KLUWER ACADEMIC PUBLISHERS GROUP
DORDRECHT / BOSTON / LANCASTER

Library of Congress Cataloging in Publication Data

Main entry under title:

Scientific rationality: The sociological turn.

(The University of Western Ontario Series in philosophy of science; v. 25)

Consists of original and revised versions of papers presented at a conference held at the University of Western Ontario, London, in 1981.

Includes bibliographies and index.

1. Science–Social aspects–Congresses. 2. Science–Philosophy–Congresses. 3. Science–History–Congresses. I. Brown, James Robert. II. Series.

Q175.4.S388 1984 500 84–15051

ISBN 90–277–1812–1

Published by D. Reidel Publishing Company,
P.O. Box 17, 3300 AA Dordrecht, Holland.

Sold and distributed in the U.S.A. and Canada
by Kluwer Academic Publishers,
190 Old Derby Street, Hingham, MA 02043, U.S.A.

In all other countries, sold and distributed
by Kluwer Academic Publishers Group,
P.O. Box 322, 3300 AH Dordrecht, Holland.

Printed in The Netherlands

For Christopher

CONTENTS

ACKNOWLEDGMENTS

A very interesting exchange took place between Larry Laudan and David Bloor in Toronto in the fall of 1980. It stimulated a conference in the spring of 1981 which I organized at the University of Western Ontario, London, while I was visiting there for the year. The initial papers in this volume are the published versions of the original Laudan-Bloor exchange. They are reprinted here with the permission of Philosophy of the Social Sciences and the authors. Most of the other papers are revised versions of what was presented at the conference in London; some are independently contributed.

I am indebted to many people for help in either organizing the conference which gave rise to this volume or in putting the volume itself together (or both). My greatest thanks go to Maxine Abrams, Robert Butts, Kathleen Okruhlik, John Nicholas, Glenn Pearce, the British Council, and, of course, the participants themselves.

James Robert Brown

INTRODUCTION: THE SOCIOLOGICAL TURN

The problem we are concerned with here is just this: How should we understand science? Are we to account for scientific knowledge[1] by appeal to the various social factors which may have been prevalent when the theory was being formulated? Should we, that is, appeal to the "interests" which a group of scientists may have had? Undoubtedly, social factors play some role, but are social causes totally responsible for the production of belief? Or should we take a different approach and account for scientific knowledge in a fashion which largely mirrors the very accounts that rational scientists themselves would have given to justify their theory choices? Perhaps we should be citing the "evidence" for the belief in question; perhaps we should be providing "good reasons" as part of the explanation for holding the belief. Which approach to understanding science is right?

The most effective modern champions of *social causation* are the members of the "Edinburgh School", a very vigorous group of sociologists and historians of science centred in the Science Studies Unit of the University of Edinburgh. They are the ones primarily responsible for the sociological turn. The principal opposition to this way of viewing science comes chiefly from philosophers as well as the more traditional sociologists and intellectual historians. But the problem we are faced with is as old as it is tricky, so, before elaborating on the modern debate, a backward glance would not be out of place.

Natural vs Cultural Sciences?

If we are properly to set the stage for the debates which take place in this volume, we could hardly do better than to start with a brief look at Karl Mannheim's view. In *Ideology*

J. R. Brown (ed.), Scientific Rationality: The Sociological Turn, 3–40.

and Utopia, the *magnum opus* of the father of the modern sociology of knowledge, Mannheim characterizes the discipline this way: "The principle thesis of the sociology of knowledge is that there are modes of thought which cannot be adequately understood as long as their social origins are obscured."[2] In itself, this seems quite innocuous; it expresses a sentiment that we all might agree to, for we all concur that *some* people believe *some* of the things they do because of social factors. Quarrels do not arise until it is a question of *which* people and *which* beliefs. Things usually become heated when it is suggested that our *scientific* beliefs have this (possibly contaminating) social origin.

There has been a long tradition in the sociology of knowledge which has sharply separated beliefs into two kinds. A boundary has been drawn between mathematics and the natural sciences on the one hand, and most everything else on the other. Included in this latter collection are such "cultural" things as religious beliefs, morals, "practical wisdom", and, often enough, the *social* sciences. The natural sciences are viewed as pristine, uncorrupted by any considerations of interest, while everything else is suspect, ideological, tarnished with subjectivity, and corrupted with naked or concealed self-interest.

Mannheim is part of this tradition (at least in some of his writings; he tended to be ambivalent on the issue). In one place he expresses the dichotomy this way:

> It may be said for formal knowledge that it is essentially accessible to all and that its content is unaffected by the individual subject and his historical-social affiliations. But, on the other hand, it is certain that there is a wide range of subject-matter which is accessible only either to certain subjects, or in certain historical periods, and which becomes apparent through the social purposes of individuals.[3]

By "formal knowledge", he means the natural sciences and mathematics. In another passage, Mannheim again gives voice to the duality between the natural and the social:

> Are the existential factors in the social process merely of peripheral significance, are they to be regarded merely as

> conditioning the origin or factual development of ideas (i.e. are they of merely genetic relevance), or do they penetrate into the "perspective" of concrete particular assertions? This is the next question we shall try to answer. The historical and social genesis of an idea would only be irrelevant to its ultimate validity if the temporal and social conditions of its emergence had no effect on its content and form. If this were the case, any two periods in the history of human knowledge would only be distinguished from one another by the fact that in the earlier period certain things were still unknown and certain errors still existed which, through later knowledge were completely corrected. This simple relationship between an earlier incomplete and a later complete period of knowledge may to a large extent be appropriate for the exact sciences (although indeed today the notion of the stability of the categorical structure of the exact sciences is, compared with the logic of classical physics, considerably shaken). For the history of the cultural sciences, however, the earlier stages are not quite so simply superseded by the later stages, and it is not so easily demonstrable that early errors have subsequently been corrected. Every epoch has its fundamentally new approach and its characteristic point of view, and consequently sees the "same" object from a new perspective.[4]

Admittedly, Mannheim hedges on the "stability of the exact sciences". Nevertheless a sharp distinction between such things as physics, chemistry, and mathematics on the one hand and what he calls the "cultural sciences" on the other hand is made.

In consequence of this dichotomy, we should have a similar dichotomy in our approach to accounting for knowledge. If we wish to explain why a certain scientist or community of scientists believes (or did believe in the past) a particular theory of physics (or some other natural science) we should look to the evidential reasons which were available. However, if it is a belief in a theological, moral, or ecomomic doctrine that is to be accounted for, then "evidence" would have little or nothing to do with it. Instead, we must look to the social factors prevalent at the time the theory choice was made which would have caused such a decision. We should point to the evidence to account for the growth of natural science; and we should point to social causes to account for every other kind of belief.

It is very important to notice, in the passage cited, why Mannheim makes the distinction between the cultural and the natural sciences. He thinks that the exact or natural sciences exemplify a "simple relationship between an earlier incomplete and a later complete period of knowledge". This reflects a view of the development of science known as the *cumulative* account. Once a fact has been discovered, it is never abandoned, additional facts are simply added toward it.

Mannheim is not alone in holding such a view of natural science nor in thinking it is a good reason for taking a hands-off attitude to it. In a recent exigetical piece on the sociology of knowledge Werner Stark voices the same sentiment when he writes, "Because man must take the facts of nature as he finds them, while the facts of culture are his own work, the social determination of knowledge will be different in the two instances."[5]

This is, however, a view of the natural sciences to which hardly anyone today, and in particular no one in this volume, gives any credence. Quite the contrary. Whether they are proponents of a "rational" or of a "sociological" approach to the understanding of science, all contributors to this volume think the natural sciences have been revolutionary, not cumulative, that many past theories, and even many past "facts" have been completely overthrown.

Indeed, attacks (especially that found in Kuhn's *The Structure of Scientific Revolutions*) on the cumulative account of science have done much to inspire the contemporary sociological turn. One of the leading figures in the Edinburgh school, Barry Barnes, puts it this way:

> It is well known that as scientific knowledge has developed, numerous mechanisms and theories have been postulated and successively set aside. This is, indeed, why so many philosophers of science have struggled to

> maintain a fact/theory distinction, and to base their justificatory rhetoric on the accumulation of facts. But there has also been a good deal of informal faith placed in the progressive quality of this sequence of theories and mechanisms. Recent historical studies, however, in particular those of T.S. Kuhn (1970), effectively undermine this faith; they demonstrate that fundamental theoretical transitions in science are not simply rational responses to increased knowledge of reality, predictable in terms of context-independent standards of inference and evaluation.[6]

The upshot, then, is that Mannheim's reason for distinguishing between the natural sciences and the cultural sciences is not such a good reason after all. The natural sciences do not develop in a cumulative fashion. But does this then mean that we must look to social causes to explain the developments in the *natural* sciences, and, indeed, to explain *all* beliefs?

The Arationality Principle

Though the dichotomy between the natural and the cultural sciences is not one which will stand up, its breakdown need not lead automatically to a full-fledged sociological account of all belief. One could maintain that what Kuhn and other historians have shown is not that the natural sciences require a sociological account, but rather that we need a new and richer notion of rationality. It will have to be a notion which, among other things, does not require a cumulative history of science. And this is just what several contemporary, post-positivistic, philosophers of science have been trying to develop.

Larry Lauldan is one prominent philosopher who is working in this direction. His *Progress and Its Problems* contains a number of interesting suggestions and proposals, but one of the most important is his denial of any sharp boundary between science and non-science (or between the natural and the cultural sciences). What is good method in science, he claims, is good method *anywhere* there are cognitive aims. Thus, physics and theology, mathematics and metaphysics, geology and sociology are all on a par as far as having the same method for *correctly* pursuing their cognitive ends. Of course, the practitioners of one or the other of these disciplines may not be following the method properly, but if they were to be completely rational they would use the same procedures. Scientific rationality simply means following the scientific method; and it is the same method for all.

As well as rejecting a demarcation between science and other cognitive activities, Laudan would also reject the idea that the practitioners of the natural sciences have made nothing but rational choices. Indeed, no one seriously believes that every scientist who ever held to some theory of physics, chemistry, or mathematics did so because of the evidence which was available. It is widely thought, even by the staunchest champions of scientific rationality, that *sometimes* a scientist will act irrationally, either in believing something which is totally crazy or in believing the right thing for the wrong reasons. And those "bad reasons" might well stem from social forces. It will simply not be true that every cognitive decision in the history of the natural sciences can be accounted for by appeal to the evidence available to those who made the decision.

In consequence, some sort of guideline is required, some sort of rule which will tell the historian how to approach individual cases in the history of science. A rule which has been proposed is this: If a belief can be explained as being the rational result of the examination of the evidence available then that should be accepted as the correct explanation. If, and only if, no such rational explanation is available, should we account for the belief by appeal to social (or other non-rational) causes. (In the jargon of "internal/external", we should only seek an external account if no internal one can be found.) Robert Merton, for instance, holds such a view: "...thought has an existential [i.e., social] basis in so far as it is not immanently [i.e., rationally] determined...."[7]

This methodological principle is explicitly adopted by Laudan who calls it the "arationality principle":

> ...basically, it amounts to the claim that *the sociology of knowledge may step in to explain beliefs if and only if those beliefs cannot be explained in terms of their*

> *rational merits*...Essentially, the arationality assumption establishes a division of labor between the historian of ideas and the sociologist of knowledge; saying, in effect, that the historian of ideas, using the machinery available to him, can explain the history of thought insofar as it is rationally well-founded and that the sociologist of knowledge steps in at precisely those points where a rational analysis of the acceptance (or rejection) of an idea fails to square with the actual situation.[8]

If a slogan is useful, perhaps Newton-Smith's will do: *Sociology is only for deviants.*[9] The sociologist is to step in *when and only when* there is some *deviation* from the norm of rationality.

It is interesting that the (traditional) sociologist Merton and the philosopher Laudan concur in their endorsement of this approach to understanding science. The recent sociological turn is as much an attack on the Mertonian way of doing the sociology of science as it is an attack on philosophers of science. The arationality principle is probably the very antithesis of the sentiments embodied in the recent sociological turn.

The Strong Programme

My aim in the balance of this introduction is to give a brief and clear, but *uncritical* exposition of the sociologists' general outlook. No attempt at a balanced presentation (between sociologists and rationalists) is made. Several of the contributors to this volume have criticisms to make, but I will let them speak for themselves below. My aim is just to set the stage.

The claims of the new cognitive sociologists of knowledge, especially of the Edinburgh school, are much stronger than the mere assertion that sometimes social factors have to be considered in order to give a complete account of an episode in the history of science. Rather it is maintained that social causes are always present; they are the determining factors. This position is known as the "Strong Programme". It is most clearly and forcefully expressed in David Bloor's important and influential work, *Knowledge and Social Imagery*.

Bloor chastises many of his fellow (traditional) sociologists for "a betrayal of their disciplinary standpoint".[10] The hands-off attitude toward science of many traditional sociologists is anathema to him. These sociologists are unnecessarily limiting the scope of their own enterprise according to Bloor; they should bring their considerable resources to bear on the very *content* of scientific knowledge. Philosophers have traditionally given sociologists only the non-natural sciences to account for, or only the irrational residue to explain. When philosophers distinguish between "discovery" and "justification", they give sociologists only the former to cope with. But, asserts Bloor, all of science is the legitimate domain of the sociologist.

Bloor has proposed the following tenets as characteristic of an adequate sociology of knowledge and a proper account of science:

1. CAUSALITY. It would be causal, that is concerned with the conditions which bring about belief or states of knowledge. Naturally there will be other types of causes apart from social ones which will cooperate in bringing about belief.
2. IMPARTIALITY. It would be impartial with respect to truth and falsity, rationality or irrationality, success or failure. Both sides of these dichotomies will require explanation.
3. SYMMETRY. It would be symmetrical in its style of explanation. The same types of cause would explain, say, true and false beliefs.
4. REFLEXIVITY. It would be reflexive. In principle its patterns of explanation would have to be applicable to sociology itself. Like the requirement of symmetry this is a response to the need to seek for general explanations. It is an obvious requirement of principle; otherwise sociology would be a standing refutation of its own theories.[11]

Bloor has much to say in defence of each of these tenets of the strong programme. The main point is that the sociologist is a scientist too, and ought to act as scientists do; he or she should try to characterize knowledge in a scientific fashion. "If sociology could not be applied in a thorough-going way to scientific knowledge it would mean" concludes Bloor, "that science could not scientifically know itself."[12] Thus, says Bloor of the sociologist:

> His ideas therefore will be in the same causal idiom as any other scientist. His concern will be to locate the regularities and general principles or processes which appear to be at work within the field of his data. His aim will be to build theories to explain these regularities. If these theories are to satisfy the requirement of maximal generality they will have to apply to both true and false beliefs, and as far as possible the same type of explanation will have to apply in both cases. The aim of physiology is to explain the organism in health and disease; the aim of mechanics is to understand machines which work and machines which fail; bridges which stand as well as those which fall. Similarly the sociologist seeks theories which explain the beliefs which are in fact found, regardless of how the investigator evaluates them.[13]

In other words, to be truly scientific, which Bloor certainly takes to be a good thing, one has to look for the *causes* of beliefs. Moreover, we do not have two theories of nature, one for explaining why a bridge stands up (when it does) and a second theory for explaining why another bridge has fallen down. We have but one theory and we use it *impartially* and *symmetrically* (with the appropriate choices of initial conditions) to explain both standing and falling bridges.

Barry Barnes adopts a methodological requirement similiar to the symmetry principle when he insists that:

> What matters is that we recognize the *sociological* equivalence of different knowledge claims. We will doubtless continue to evaluate beliefs differently ourselves, but such evaluations must be recognized as having no relevance to the task of sociological explanation; as a methodological principle we must not allow our evaluation of beliefs to determine which form of sociological account we put forward to explain them.[14]

A common argument directed against such a full-blooded sociology of scientific knowledge goes like this: If all belief is caused by social factors then this must be true of the strong programme as well. Therefore, the strong programme falls into a self-refuting relativism; it undermines its own position. Bloor, however, thinks this is not damaging in the least. The principle of *reflexivity* is simply an admission of the premiss of this argument. But the conclulsion does not follow, says Bloor, unless social determination implies falsehood. But causation does not imply error; so the charge of "self-refutation", he maintains, will not stick.

The requirement of being "scientific" is not the only motivation for the recent sociological turn. Another factor is the perceived success of recent case studies. The last decade has seen a great number of new sociological accounts of old stories and I shall summarize three of these below for handy reference.

A third factor which has played a big role in the sociological turn stems from recent (largely philosophical[15]) work on "underdetermination" and "incommensurability". The problem of underdetermination, briefly, is this: there are indefinitely many logically possible theories which are compatible with the empirical data. Thus, the experimental evidence cannot pick out one from among these as being the uniquely correct or true theory. So, it is concluded, the decision which scientists do make cannot be based on rational considerations and we must look elsewhere for an explanation of the choice. As Bloor puts it, "The theoretical component of knowledge is a social component".[16] The philosopher Mary Hesse is sympathetic; she comes to the same conclusion in her new work *Revolutions and Reconstructions in the Philosophy of Science*.[17] Sociological explanations will have to be given since the evidence, she claims, will not determine one theory as being better than its rivals.

The work on incommensurability by Kuhn, Feyerabend, and Wittgenstein has had a great deal of influence on Barnes. In his new book on Kuhn he remarks, "Nothing in the nature of things, or the nature of language, or the nature of past usage, determines how we employ, or correctly employ our terms."[18]

The determining factors in how we use our concepts, and especially in how we extend them, Barnes claims, are social factors.

Some Examples

I have been describing some of the programmatic features of the Edinburgh school's view of science, but now, in order to help make clearer just what the recent sociological turn is, a few examples are in order. Brief descriptions of some paradigmatic case studies will throw considerable light on the new sociological enterprise. The ones which I have chosen to recount are from the collection of examples that Edinburgh school members have either produced themselves or have repeatedly referred to in approving terms.

This volume contains only sketches of a few case studies (with the notable exception of Robert Butts's long study of Kant and ESP). Yet clearly, concrete examples are important to the issue. No single case study will make or break either side, but, I think all contributors to this volume will readily agree that historical considerations have some role or other to play in deciding the issure between sociologists and rationalists. Just what that role is, is hard to say. It cannot be a case of simple induction since the proponents of the arationality principle can readily allow that there have been *some* instances of "bad" science and that consequently there should be cases where the rationalist and the sociologist concur on the cause (i.e., sociological) of the scientific belief in question. Whatever the evidential role these examples play, legitimate or not, there can be no doubt that they have been influential in discussions of the issue. Here, for handy reference, are brief synopses of three of the most influential among them. After each, I cite a reply which is highly critical of the study in question. (My uncritical expositions of these three case studies should not be taken as endorsements.)

(1) *Forman on Weimar Culture and Causality.* The scientists of the Weimar Republic, according to Paul Forman,[19] perceived themselves to be under attack. And their perceptions were correct for they were indeed living in a hostile intellectual

environment. Following the war and Germany's collapse, the public was quite disillusioned with science and technology. The general tenor of the times was mystical and antirational. Even the little bits of science that the public was interested in, such as relativity, were often used to support anti-rationalist positions. The mystical and romantic public sentiments of the post-war German public were, according to Forman, in direct opposition to the perceived spirit of science, which in turn was seen as mechanistic, rationalistic, and causalistic.

This sense of general intellectual crisis was epitomized by Spengler's *Decline of the West*. The content of Western mathematics and physics, according to Spengler, expresses the "Faustian" nature of contemporary Western culture; and the essential ingredient is the *Kausalitatsprinzip*. Mathematics, physics, causality, and rationality are lumped together and linked to death. In oppositioin to this is the creative, the living, that which embraces the incomprehensible "Destiny" *(Lebensphilosophie)*. But Spengler also added that physics in his day had exhausted its possibilities; doubts were arising about its principal concepts. Salvation for western science would come about; he thought, but only when that science returned to its *spiritual* home.

Various remarks from a number of prominent scientists who embraced or at least seriously alluded to the popular sentiment, are cited by Forman. Wien, von Mises, Weyl, Born, Sommerfelde, and others are noted as making significant concessions to the importance of "spiritual values", "the mystery of things", "traditional German idealism", etc. In fact, these concessions went on to such an extent that Forman calls it a "capitulation to Spenglarism".[20]

The then popular view of a *crisis of culture* precipitated the crises in the various sciences after the war. As for physics in particular, "The *possibility* of the crisis of the old quantum theory was", says Forman, "dependent upon the physicists' own craving for crises, arising from participation in, and adaptation to, the Weimar intellectual milieu".[21]

Forman characterises all those who embraced a nondeterministic view as espousing it in a moralistic and almost religious fashion, and moreover, as doing so only after the most frivolous consideration. In short, there was nothing rational about it. Forman concludes:

> ...[S]uddenly deprived by a change in public values of the approbation and prestige which they had enjoyed before and during World War I, the German physicists were impelled to alter their ideology and even the content of their science in order to recover a favorable public image. In particular, many resolved that one way or another, they must rid themselves of the albatross of causality.
>
> In support of this general interpretation I illustrated and emphasized the fact that the program of dispensing with causality in physics was, on the one hand, advanced quite suddenly *after* 1918 and, on the other hand, that it achieved a very substantial following among German physicists *before* it was "justified" by the advent of a fundamentally acausal quantum mechanics. I contended, moreover, that the scientific context and content, the form and level of exposition, the social occasions and the chosen vehicles for publication of manifestoes against causality, all point inescapably to the conclusion that substantive problems in atomic physics played only a secondary role in the genesis of this acausal persuasion, that the most important factor was the social-intellectual pressure exerted upon the physicists as members of the German academic community.[22]

And, moreover, he adds,

> Although a readiness to view atomic processes as involving a "failure of causality" proved to be, and remains, a most fruitful approach, before the introduction of a rational acausal quantum mechanics the movement to dispense with causality expressed less a research program than a proposal to sacrifice physics, indeed the scientific enterprise, to the *Zeitgeist*.[23]

For a strong critique of Forman see John Hendry (1980), "Weimar Culture and Quantum Causality".[24]

(2) *Shapin on the Edinburgh Phrenology Debates.* A Viennese

doctor, Franz Joseph Gall (1758-1828), was the founder of phrenology. Its main principles were that the organ of the mind is the brain; that the brain is composed of separate organs, each associated with a different mental faculty; and that the size of the organ is related to the power of the associated mental faculty.

The phrenologists encountered two hostile traditions in Edinburgh: one was Anatomy, the other was Philosophy (the tradition from Thomas Reid and Dugal Stewart to William Hamilton). The ensuing debates in the early years of the nineteenth century were particularly strenuous. Steven Shapin has examined these debates and has given a sociological account of the episode which runs as follows.[25]

The early decades of the nineteeth century saw a serious growth of tension between the social classes in Edinburgh. The mercantile classes began to resent and reject the values and privileges of the upper classes, the landed gentry, the lawyers, and the professionals. The emerging middle classes established a number of institutions for their own purposes, inclulding their own paper, *The Scotsman*, which, says Shapin,[26] "...was critical of the Tories, the University, the established Church, and what it saw as intellectual obscurantism". And he adds, "...it was no coincidence that *The Scotsman* supported the phrenologists in their dispute with the moral philosophers".

Shapin begins by noting the enormous enthusiasm for phrenology amongst the middle and working classes. Popular lectures on the subject were amazingly well-attended. Working class institutions regularly offered courses on the subject and even gave it pride of place. The attitude of the University, however, was quite different. Phrenology was never taught there. Often *anti*-phrenology lectures were given at the University but phrenologists were seldom or never allowed to reply. The membership of the very elite Royal Society of Edinburgh and the membership of the Phrenological Society were virtually disjoint. And so, says Shapin, the phrenology supporters must be considered "outsiders".

Shapin further claims that:

> ...British phrenology was a social reformist movement of the greatest significance. Combe (George Combe, the

> most prominent phrenologist in Edinburgh) and his circle vigorously, and to some extent successfully, agitated for penal reform, more enlightened treatment of the insane, the provision of scientific education for the working classes, the education of women, the modification of capital punishment laws and the rethinking of British Colonial Policy.[27]

He certainly admits that there was a *technical* debate with all the trappings of regular science, that is, argument, evidence, and so on; nevertheless "...to say there *was* a technical debate is not to say that it can be or was separated from the social conflict, nor that such a technical debate does not reflect social and institutional divisions."[28]

Shapin's account of the Edinburgh phrenology debates is part of a debate he himself had with G. Cantor.[29] The Cantor-Shapin confrontation especially repays close examination because it is a head-on clash of two diametrically opposed ways of understanding science. While Shapin accounts for the episode in terms of social factors, Cantor, who is much more of a traditional intellectual historian, looks to the evidence, the arguments, and the reasons which were adduced for each side.

Cantor criticizes Shapin's sociological account for a number of reasons. On the one hand he disputes some of the details such as Shapin's claim that there was an increase of social tension at the time; he also notes that the overlap of the memberships of the "elite" Royal Society of Edinburgh and the Phrenological Society was not negligible as Shapin claims. But Cantor's other charges are of a more general programmatic and philosophical nature: What is the relation between the content of the beliefs and social factors? Shapin has not spelled out the connection. Why should one look, as Shapin demands, to social conflict in order to explain intellectual conflict? The main complaint coming from Cantor, however, is that Shapin has failed to pay attention to the reasoning processes that went on and to the evidence which was available. In other words, he has failed to take the *intellectual content* of the Edinburgh phrenology debates seriously.

(3) *Farley and Geison on the Politics of Spontanious Generation.* Louis Pasteur and Felix Pouchet had a famous debate over

spontaneous generation which lasted for about five years during the middle of the nineteenth century. Contrary to traditional accounts of this famous episode in the history of science, John Farley and Gerald Geison "believe that (their) reexamination of the Pasteur-Pouchet debate reveals the direct influence of extrinsic factors on the conceptual content of serious science".[30]

Spontaneous generation, in a nutshell, is the doctrine that living organisms can arise independently, without parents, from either inorganic matter (abiogenesis) or from organic debris (heterogenesis). Typical historical accounts claim that Pouchet set out with preconceived ideas favouring spontaneous generation while Pasteur, with his flawless experimental technique, destroyed the doctrine by doing a series of conclusive tests. It is quite a different picture that Farley and Geison present.

In the middle of the nineteenth centruy, France was a politically conservative, even reactionary, country. Louis Napoleon had come to power in 1848 with the support of the Catholic Church; and religious and political issues were inseparable. Church and state faced perceived common enemies: republicanism and atheism. Indeed, very often atheists, positivists, and materialists truly were opposed to both church and state. (One striking instance is Clemence Royer who translated Darwin's *Origin* in 1864. In her preface she explicitly attacked the Catholic Church calling it corrupt, ignorant, and responsible for all societal ills.) Generally, there were considerable social tensions, and these were reflected in people's attitudes toward science.

The Pasteur-Pouchet debate can be said to begin with the appearance in 1859 of Pouchet's *Heterogenie, ou traite de la generation spontanee*. Aware of the political climate that he was working in, Pouchet made explicit disclaimers of atheism and offered a prolonged justification of his claim that his account of the theory of spontaneous generation was perfectly compatible with orthodox science *and* orthodox religion. His version of the theory posited a "plastic force" which was capable of organizing molecules in special ways and endowing them with vitality. It was an egg, not an adult organism, which was spontaneously generated in this way. In defending himself against possible religious challenges, Pouchet claimed that scripture did not contradict his claim that God might be always creating life; there is no reason to think that he stopped after the sixth day.

Pasteur's background to the debate is characterised by curiously contradictory trends, according to Farley and Geison. On the one hand, Pasteur had done a great deal of work on fermentation. In his work he needed to face the issue of the origin of the organisms responsible for the process. Pasteur was arguing against any chemical theory of fermentation, and so he had to argue that the organisms pre-existed and did not arise heterogenetically. His later attack on Pouchet would seem to be of a piece with his views on fermentation.

However, Pasteur also did a great deal of work in crystallography where he had become convinced that molecular asymmetry (which manifested itself in optical phenomena) was intimately connected to life. Pasteur speculated that the force which brought molecular asymmetry about, an asymmetric force, was a kind of ordinary physical force. All of this, of course, suggests that abiogenesis could occur under ordinary mechanistic circumstances. Indeed, in his laboratory Pasteur tried to "imitate nature" and to "introduce asymmetry into chemical phenomena".[31] Concerning all of this, Farley and Geison remark:

> In fact, of course, Pasteur did not succeed in creating asymmetry or life and temporarily abandoned these experiments. But he continued to believe that abiogenesis should be possible under some such experimental conditions. He thus came into the debate over spontaneous generation faced with a curious dilemma. On the one hand, his work on fermentation led him to discount the possibility of heterogenesis, while on the other his theoretical views on asymmetry and life led him not only to believe in the possibility of abiogenesis but actually to attempt such a feat experimentally. If it seems illogical simultaneously to believe that life can be produced artificially from inorganic elements but not from a rich organic soup, it is essential to recall that Pasteur reached this paradoxical position as the result of two quite separate research problems and to emphasize the

> distinction in his mind between symmetric chemical influences and asymmetric physical forces. Nevertheless, and this is the central point, Pasteur could deny the possibility of spontaneous generation only by suppressing part of his own scientific beliefs.[32]

Pasteur's political views were in complete harmony with the orthodoxies of the Second Empire, which is to say he was very conservative. He was a strong supporter of Louis Napoleon, dedicating a book to the Emperor and another to the Empress. And in turn he benefited greatly from Imperial favour. He once ran for the senate as a conservative and as a champion of the established order. Pasteur, according to Farley and Geison, had a "general preference for order and stability over free speech, civil liberty or even democracy, whose potential for anarchy and mediocrity he feared".[33] It was this strong conservative political stance which influenced Pasteur to take the scientific stance that he adopted. In conclusion, Farley and Geison say:

> Remarkably enough, we are led to a conclusion precisely the opposite of that usually attached to the Pasteur-Pouchet debate. For we are persuaded that external factors influenced Pasteur's research and scientific judgement more powerfully than they did the defeated Pouchet. Having formulated his version of spontaneous generation prior to the politically significant Darwinian controversy in France, Pouchet maintained his views with striking consistency in spite of their presumed threat to orthodox religious and political beliefs which he fully shared. By contrast, Pasteur's public posture on the issue seems to reveal a quite high degree of sensitivity to reigning socio-political orthdoxies.[34]

This cases study has been criticised by Nils Roll-Hanson (1979) who claims external factors counted for little in the debate.[35]

There are a number of other examples which I could have just as easily given, for instance, the studies by T. Brown, D. MacKenzie, H. Collins, M. Mulkay, or any of those appearing in Barnes and Shapin (eds. 1979). (See the bibliography for details.) They are recommended as further examples of the sociological approach to understanding science.

The Ongoing Debate

Since Plato first took up the cudgels against Protagoras when the latter said that man is the measure of all things, the battle over cognitive relativism has been fought and fought again. But it would be a mistake, I think, to assume that nothing is ever achieved. Each version of the debate has its own peculiarities, and subtle new points often come to light in each clash. This is certainly true of the present debate; the Edinburgh school is unique in many ways. For example, unlike any of their precursors, they are not out to "expose" science. They are unabashedly pro-science. Bloor, for example, says he is trying to be "scientific" in attempting a "science of science."

Whether he succeeds in providing a scientific account of science is a matter of some import. Larry Laudan, in the first paper in this volume, "The Pseudo-Science of Science?", thinks he fails completely; and several of the other other contributors to this volume share Laudan's pessimism on this score, (though often to a lesser degree.) The main focus of Laudan's attack on the strong programme is the symmetry principle. All the other principles, if properly interpreted, he claims, are rendered innocuously true. It is the symmetry principle, the principle which says that all explanations must be the same, whether of true or false, rational or irrational beliefs, which is the heart of Bloor's programme. Laudan takes the strongest possible exception to this, claiming that this is not the way science works; and he provides examples to illustrate his point. The main claim, perhaps, is just this: Reasons are a kind of cause, and sometimes these reason-causes are present. They are not always present, but they are not always absent either. Consequently, the symmetry principle, which is an all-or-nothing principle, must be wrong.

In his rejoinder, "The Strengths of the Strong Programme", David Bloor replies with a number of examples to illustrate his point of view. He also stresses the problem of the underdetermination of theories by data, claiming that Laudan fails to take the matter seriously, and so begs many large questions.

The opening salvos by Laudan and by Bloor provide the main point of departure for most of the remaining contributions. Gary Gutting provides a considerable amount of clarificatory analysis in "The Strong Programme: A Dialogue". In particular, he distinguishes between two types of explanation of belief and thereby sheds some light on what is, and is not, really at issue in the debate between philosophers and sociologists.

In his contribution, "Problems of Intelligibility and Paradigm Instances", Barry Barnes worries about antagonists, such as Bloor and Laudan, talking past one another. He is somewhat optimistic, however, that a few paradigm case studies should go a long way toward clarifying and then settling the debate.

Ernan McMullin calls for a middle course in "The Rational and the Social in the History of Science". He sees Laudan and Bloor as both being extremists. Presumptions of unrestricted rationality or unrestricted social causation are both unwarranted, he says.

In "A Plague on Both Your Houses", Ian Jarvie, like some other contributors to this volume, wants to set out on an entirely different course. Laudan and Bloor both accept the internal/external distinction (with Laudan claiming that most science falls under the first heading and Bloor claiming that it all falls under the second). The problem, as Jarvie sees it, is with the distinction itself. Give that up, he says, and we will have quite a different view of science from that presented by either Laudan or Bloor.

Andrew Lugg's "Two Historiographic Strategies" considers various aspects of the sociological turn and counters several of the standard arguments put forward for the Edinburgh school's way of approaching history. He largely favours the methods and techniques of the traditional historians of science.

In the only paper in this volume to provide a detailed case study, Robert Butts sides mainly with Laudan and the rationalists. As reason for doing so, he cites the interesting and unusual case of Kant and ESP. In his analysis he is critical of an earlier sociological attempt to account for Kant's actions. Instead, using machinery developed in the Gutting article, he concludes on the side of the rationalists.

Jerry Gaston finds himself in sympathy with small points that both Laudan and Bloor make. However, in "On the Sociology of Belief, Knowledge, and Science" he does not expect any ultimate reconciliation of the warring parties. He also protests against the strong programmers' claim that "Mertonian" sociologists, such as himself, are against a *cognitive* sociology of science. Not so, he says, they just have not got around to doing it yet.

In "Scientific and Other Interests", John Nicholas makes an important point: Suppose theories are underdetermined. If this is a problem for cognitive aims, then it must also be a problem for other aims as well. There will not be a right answer to the question, How do we best pursue our interests? Nicholas goes on to claim that we ought to drop the whole idea of inductive logic and replace it with *inductive behavior*, a change he sees as being congenial to an interests model of scientific thinking. The volume ends with Bloor's replies.

I think it is safe to say two things at this point: one is that the contributions to this volume carry the debate considerably forward. The second is that there is still a long way to go. There are a number of important questions which are still up in the air, and an overall settlement would require putting these matters to rest. They include such questions as this: What is rationality? Participants in the debate work with concepts which are either vague or mutually conflicting. For instance, Laudan and Popper would find themselves on the same side against the strong programme; they would both champion scientific rationality. Yet what each means by "rationality" is quite different from what the other means. What impact, if any, does this have on the rationalists' case against cognitive sociologists?

Another question concerns instrumental rationality. Is there such a thing. Are there recognizably better ways of attaining one's non-cognitive goals? If there are, then why could these same methods not be used for cognitive ends as well? And if not, then how does the sociologist explain *any* behavior?

Finally, how do case studies work? Both rationalists and sociologists use individual episodes in the history of science to illustrate their respective positions. The case studies seem to fulfill this function very well. However, both sides claim more; not only do case studies illustrate, but they also are supposed to serve as a sort of *evidence*. If this is so, then *how* do they function as evidence? No individual case study could serve as a crucial experiment, for it is conceded by the rationalist that *some* episodes in the history of science were not rational. Thus, some would require a sociological (or other external) explanation. So how then does a case study support one side or the other?

These are just a few of the many real problems which remain open, problems which have a definite bearing on the outcome of the debate over how best to understand science. There has been a growing interest in these sorts of issues in recent years, and this is a turn of events we can only welcome. If a greater number of historians, philosophers, and sociologists work on these problems, then so much greater will be our eventual understanding of science.

Department of Philosophy
University of Toronto

Notes

1. Philosophers typically use "knowledge" to mean something like *true justified belief*. Accordingly, the sociology of knowledge might better be called "the sociology of belief". However, I shall follow the sociologists' custom of using "knowledge" and "belief" interchangably.
2. Mannheim, *Ideology and Utopia*, Routledge and Kegan Paul, 1936.
3. *Ibid*.., p. 150.
4. *Ibid*., p. 243.
5. Stark, "The Sociology of Knowledge," Edwards (ed.) The *Encyclopedia of Philosophy*, vol. 7, Collier Macmillan, 1967, p. 477.
6. Barnes, *Interests and the Growth of Knowledge*, Routledge and Kegan Paul, 1977, p. 23. Kuhn has been very influential in this regard; so much so that Barnes has just written a new book, *T.S. Kuhn and Social Science*, which uses Kuhn to justify a sociological approach to science.
7. Merton, *Social Theory and Social Structure*, Freepress,

1968, p. 516.

8. Laudan, *Progress and Its Problems,* University of California Press, 1977, p. 202.
9. Newton-Smith, *The Rationality of Science,* Routledge and Kegan Paul, 1981, p. 238.
10. Bloor, *Knowledge and Social Imagery,* Routledge and Kegan Paul, 1976, p. 1.
11. *Ibid.,* p. 5
12. *Ibid,* p. 40.
13. *Ibid.,* p. 3.
14. Barnes, *Interests and the Growth of Knowledge,* Routledge and Kegan Paul, 1977, p. 25.
15. See, for example, Feyerabend, 1975; Kuhn, 1970; Quine, 1953, 1960.
16. Bloor, *loc. cite.,* p. 13.
17. See especially chapter 2.
18. Barnes, *T. S. Kuhn and Social Science,* p. 30. See also Barnes 1980 for an elaboration of this theme.
19. Forman, "Weimar Culture and Causality" in *Historical Studies in the Physical Sciences,* McCormmich (ed.), 1971.
20. *Ibid.,* p. 55.
21. *Ibid.,* p. 62.
22. *Ibid.,* p. 109f.
23. *Ibid.,* p. 112f.
24. J. Hendry, "Weimar Culture and Quantum Causality", *History of Science,* xviii (1980).
25. S. Shapin, "Phrenological Knowledge and the Social Structure of Early Nineteenth-Century Edinburgh," *Annals of Science,* 32 (1975), pp. 219-243.
26. *Ibid.,* p. 224.
27. *Ibid.,* p. 232.
28. *Ibid.,* p. 234.
29. G. Cantor, "Phrenology in Early Nineteenth-Century Edinburgh: An Historiographical Discussion," *Annals of Science,* 32 (1975), pp. 195-218 and "A Critique of Shapin's Social Interpretation of the Edinburgh Phrenology Debate," *Annals of Science,* 33 (1975), pp. 245-256.
30. Farley and Geison, "Science, Politics and Spontaneous Generation in Nineteenth-Century France: The Pasteur-Pouchet Debate," *Bulletin of the History of*

Medicine, 48, 1974, pp. 161-198. Passage quoted from p. 162.

31. *Ibid.*, p. 178.
32. *Ibid.*, p. 178f.
33. *Ibid.*, p. 187.
34. *Ibid.*, p. 197.
35. Nils Roll-Hansen (1979) "Experimental Method and Spontanious Generation: The Controversy between Pasteur and Pouchet, 1859-64. *Journal of the History of Medicine*, 34, pp. 273-292.

Bibliography

This bibliography is not exhaustive. Much more can be found in, for instance, Barnes and Edge (1982). It does contain most of the major writings of some members of the Edinburgh School, i.e., Barry Barnes, David Bloor, and Steven Shapin. I thank them for sending me the relevant information. This bibliography consists mainly of sociological works. It does not attempt a "balanced" presentation; only a few rationalist works are cited.

Agassi, J. (1963), "Towards an historiography of science', *History and Theory*, vol. 2, pp. 1-23.

Barber, B. (1952), *Science and the Social Order*. New York: Free Press.

Barber, B. (1961), "Resistance by scientists to scientific discovery", *Science*, vol. 134, no. 3479, pp. 596-602.

Barber, B. and Fox, R. (1958), "The case of the floppy-eared rabbits", *American Journal of Sociology*, no. 64, pp. 128-36.

Barber, B. and Hirsch, W. (eds.) (1962), *The Sociology of Science*. New York: Free Press.

Barnes, S. B. (1969), "Paradigms: scientific and social", *Man*, vol. 4, pp. 94-102.

Barnes, S.B. (1971), "Making out in industrial research", *Science Studies*, vol. 1, no. 2, pp. 157-75.

Barnes, S.B. (ed.) (1972), *Sociology of Science*. London: Penguin.

Barnes, S.B. (1972a), "On the reception of scientific beliefs", in Barnes (1972).

Barnes, S.B. (1972b), "Sociological explanation and natural science", *European Journal of Sociology*, vol. 13, no. 3, pp.

373-93.

Barnes, S.B. (1973), "The comparison of belief systems: Anomaly versus falsehood", in Horton and Finnegan, (eds.) *Modes of Thought*. London: Faber.

Barnes, S.B. (1974), *Scientific Knowledge and Sociological Theory*. London: Routledge and Kegan Paul.

Barnes, S.B. (1976), "Natural rationality: A neglected concept in the social sciences", *Philosophy of the Social Sciences,* vol. 6, no. 2, pp. 115-26.

Barnes, S.B. (1977c), *Interests and the Growth of Knowledge*. London: Routledge and Kegan Paul.

Barnes, S.B. (1979), Vicissitudes of belief", (Review of Laudan (1977)), *Social Studies of Science,* vol. 9, pp. 247-63.

Barnes, S.B. (1980), "On the causal explanation of scientific judgement", *Social Science Information,* vol. 19, pp. 685-95.

Barnes, S.B. (1981), "On the conventional character of knowledge and cognition", *Philosophy of the Social Sciences,* vol. 11, no. 3, pp. 303-33.

Barnes, S.B. (1981a), "On the 'hows' and 'whys' of cultural change", *Social Studies of Science,* vol. 11, no. 4, pp. 481-98.

Barnes, S.B. (1982), *T.S. Kuhn and Social Science*. New York: Columbia University Press.

Barnes, S.B. (1982a), "On the extensions of concepts and the growth of knowledge", *Sociological Review,* vol. 30, no. 1, pp. 23-44.

Barnes, S.B. (1982b), "The science-technology relationship" A model and a query", *Social Studies of Science,* vol. 12, no. 1, pp. 166-72.

Barnes, S.B. and Bloor, D. (1982), "Relativism, rationalism, and the sociology of knowledge", in M. Hollis and S. Lukes (eds.) *Rationality and Relativism*. Oxford: Blackwell

Barnes, S.B. and Dolby, R.G.A. (1970), "The scientific ethos: A deviant viewpoint", *European J. Sociology,* vol. 11, no. 1, pp. 3-25.

Barnes, S.B. and Edge, D. (1982), *Science in Context: Readings in the Sociology of Science*. Boston: MIT Press.

Barnes, S.B. and Law, J. (1976), "Areas of ignorance in normal science", *Sociological Review,* vol. 24, no. 1, pp. 115-24.

Barnes, S.B. and Law, J. (1976a), "Whatever should be done with indexical expressions?", *Theory and Society,* vol. 3, no. 2, pp. 223-37.

Barnes, S.B. and MacKenzie, D. (1975), "Biometriker versus Mendelianer. Eine kontroverse und ihre erklarung', Kolner *Zeitschrift fur Soziologie und Sozialpsychologie,* special issue 18, pp. 165-96.

Barnes, S.B. and MacKenzie, D. (1979), "On the role of interests in scientific change", *Sociological Review Monographs,* (Special Issue on "Rejected Knowledge")

Barnes, S.B. and MacKenzie, D. (1979a), "Scientific judgment in the Biometry-Mendelism controversy", in Barnes and Shapin (eds.) *Natural order.*

Barnes, S.B. and Shapin, S. (1976), "Head and hand: rhetorical resources in British pedagogical writing, 1770-1850", *Oxford Review of Education,* vol. 2, no. 3, pp. 231-54.

Barnes, S.B. and Shapin, S. (1977), "Science, nature and control: Interpreting mechanics' institutes", *Social Studies of Science,* vol. 7, no. 1, pp. 31-74.

Barnes, S.B. and Shapin, S. (1977a), "Where is the edge of objectivity?, (Essay Review), *British J. Hist. Sci.,* vol. 10, no. 1, pp. 61-66.

Barnes, S. B. and Shapin, S. (eds.) (1979), *Natural Order.* London: Sage.

Barnes, S.B. and Shapin, S. (1979a), "Darwin and social Darwinism: Purity and History", in Barnes and Shapin (eds.) *Natural Order.*

Ben-David, J. (1968), *Fundamental Research and the Universities.* Paris: OECD.

Ben-David, J. (1971), *The Scientist's Role in Society.* Englewood Cliffs, N.J.: Prentice-Hall.

Ben-David, J. (1977), "Organization, social control, and cognitive change in science", in J. Ben-David and T.N. Clarke (eds.) *Culture and Its Creators.* Chicago and London: University of Chicago Press.

Berger, P. and Luckmann, T. (1967), *The Social Construction of Reality.* London: Allen Lane.

Bloor, D. (1970), "Is the official theory of mind absurd?", *British Journal for the Philosophy of Science,* vol. 21, pp. 167-83.

Bloor, D. (1970a), "Explanation and analysis in Strawson's 'Persons'," *Australian Journal of Philosophy,* vol. 48, no. 1, pp. 2-9.

Bloor, D. (1971), "The dialectics of metaphor", *Inquiry*, vol. 14, pp. 430-44.

Bloor, D. (1971a), Two paradigms for scientific knolwedge?", *Science Studies*, vol. 1, pp. 101-115.

Bloor, D. (1973), "Wittgenstein and Mannheim on the sociology of mathematics", *Studies in History and Philosophy of Science*, vol. 4, no. 2, pp. 173-91.

Bloor, D. (1973a), "Are philosophers averse to science?", in D. Edge and J. Wolfe (eds.) *Meaning and Control*. London: Tavistock, pp. 1-17.

Bloor, D. (1974), "Popper's mystification of objective knowledge", *Science Studies*, vol. 4, pp. 65-76.

Bloor, D. (1974a), "Rearguard rationalism", *Isis*, vol. 65, pp. 249-53.

Bloor, D. (1975), "Psychology or epistemology", *Studies in History and Philosophy of Science*, vol. 6, pp. 382-95.

Bloor, D. (1976), *Knowledge and Social Imagery*. London: Routledge and Kegan Paul, 156 pp.

Bloor, D. (1978), "Polyhedra and the abominations of Leviticus", *British Journal for the History of Science*, vol. 11, pp. 245-72.

Bloor, D. (1979), "Reply to Worrall", *British Journal for the History of Science*, vol. 12, pp. 78-80.

Bloor, D. (1980), "Comments on Hesse's Work on Inductive Logic", in L. Cohen and M. Hesse (eds.) *Applications of Inductive Logic*. Oxford: Clarendon Press, pp. 236-40.

Bloor, D. (1980a), "Klassifikation und wissenssoziologie: Durkheim und Mauss Neu Betrachtet", *Kolner Zeitschrift fur Soziologie und Sozialpsychologie*, Sonderheft 22, Wissenssoziologie, pp. 20-51.

Bloor, d. (1981), "The strengths of the strong programme", *The Philosophy of the Social Sciences*, vol. 11, pp. 199-213. [Reprinted in this volume.]

Bloor, D. (1981a), "Hamilton and Peacock on the essence of algebra", in H. Mehrtens, H. Bos, I Schneider (eds) *Social History of Nineteenth-Century Mathematics*. Boston, Basel and Stuttgart: Birkhauser, pp. 202-32.

Bloor, D. (1981b), Entries in W. Bynum, R. Porter and E.J. Browne (eds.), *Dictionary of the History of Science*. London:

Macmillan: 'Anomaly', 'Grid-group Analysis', 'Paradigm', 'Rational Reconstruction', 'Relativism', 'Revolutionary Science', 'Sociology of Scientific Knowledge'.

Bloor, D. (1982), "Durkheim and Mauss revisited: Classification and the sociology of knowledge", in *Studies in History and Philosophy of Science,* vol. 13, no. 4, pp. 267-297.

Bloor, D. (1983), *Wittgenstein: A Social Theory of Knolwedge.* London: Macmillan.

Bloor, D. (forthcoming), "Sociological aspects of the history of matter theory, 1850-1930", in R. Harre (ed.) *The History of Matter Theory.* Oxford.

Bloor, D. and Barnes, B. (1982), "Rationalism, relativism and the sociology of knowledge", in M. Hollis and S. Lukes (eds.) *Rationality and Relativism.* Oxford: Blackwell.

Bloor, D. and Bloor, C. (1982), "Twenty industrial scientists: A preliminary exercise", in M. Douglas (ed.) *Essays in the Sociology of Perception.* London: Routledge and Kegan Paul, pp. 83-102.

Brannigan, A. (1981), *The Social Basis of Scientific Discoveries.* Cambridge.

Brown, T.M. (1974), "From mechanism to vitalism in eighteenth-century English physiology", *Journal of the History of Biology,* vol. 7, pp. 179-216.

Cantor, G.N. (1975), "Phrenology in early nineteenth-century Edinburgh: An historiographical discussion", *Annals of Science,* 32, pp. 195-218.

Cantor, G.N. (1975a), "A critique of Shapin's social interpretation of the Edinburgh phrenology debate", *Annals of Science,* 33, pp. 245-256.

Cole, J.R. and Cole, S. (1973), *Social Stratification in Science.* Chicago and London: University of Chicago Press.

Collins, H.M. (1974), "The TEA set: Tacit knowledge and scientific networks", *Science Studies,* vol. 4, pp. 165-85.

Collins, H.M. (1975), "The seven sexes: A study in the sociology of a phenomenon, or the replication of experiments in physics", *Sociology,* vol. 9, pp. 205-24.

Collins, H.M. (1981), "The social destruction of gravitational radiation", *Social Studies of Science,* vol. 11, no. 1, pp. 33-62.

Collins, H.M. and Pinch, T.J. (1979), "The construction of the paranormal: Nothing unscientific is happening", in Wallis (ed.) *On the Margins of Science*, pp. 237-70.

Collins, H.M. and Pinch, T.J. (1982), *Frames of Meaning*. London: Routledge and Kegan Paul.

Coleman, W. (1970), "Bateson and chromosomes: Conservative thought in science", *Centaurus*, vol. 15, no. 3-4, pp. 228-314.

Crane, D. (1972), *Invisible Colleges*. Chicago: University of Chicago Press.

Curtis, J.E. and Petras, J.W. (1970), *The Sociology of Knowledge*. New York: Praeger.

Douglas, Mary (1966), *Purity and Danger: An Analysis of Concepts of Pollution and Taboo*. London: Routledge and Kegan Paul.

Douglas, Mary (1970), *Natural Symbols*. London: Barrie and Jenkins.

Duhem, P. (1962), *The Aim and Structure of Physical Theory*. New York: Atheneum.

Durkheim, E. (1915), *The Elementary Forms of the Religious Life*. London: Allen & Unwin.

Durkheim, E. (1938), *The Rules of Sociological Method*. Glencoe, Ill.: The Free Press.

Durkheim, E. and Mauss, M. (1903), *Primitive Classification*. Eng. trans. (1963). London: Cohen and West.

Easlea, B. (1973), *Liberation and the Aims of Science*. London: Chatto and Windus.

Edge, D.O. and Mulkay, M.J. (1976), *Astronomy Transformed*. New York: Wiley Interscience.

Elias, N. (1971), "Sociology of knowledge: New perspectives", *Sociology*, vol. 5, pp. 149-68 and 355-70.

Elkana, Y. (1974), *The Discovery of the Conservation of Energy*. London: Hutchinson.

Farley, J. (1977), *The Spontaneous Generation Controversy from Descartes to Oparin*. Baltimore: Johns Hopkins University Press.

Farley, J. and Geison (1974), "Science, politics and spontaneous generation in nineteenth-century France: The Pasteur-Pouchet debate", *Bulletin of the History of Medicine*, 48, pp. 161-198.

Feuer, L.S. (1971), "The social roots of Einstein's theory of relativity", *Annals of Science,* vol. 27, pp. 277-98, 313-43.

Feuer, L.S. (1974) *Eistein and the Generations of Science.* New York: Basic Books.

Feyerabend, P. (1975). *Against Method.* London: NLB.

Fleck, L. (1935), *Entstehung und Entwicklung einer Wissenschaftliche Tatsache* (English translation F. Bradley and T.J. Trenn). Published as *Genesis and Development of a Scientific Fact,* University of Chicago Press, in 1979.

Forman, P. (1971), "Weimar culture, causallity and quantum theory, 1918-1927: Adaptation by German physicists and mathematicians to á hostile intellectual environment", in R. McCormmach (ed.) *Historical Studies in the Physical Sciences,* No. 3. Philadephia, Penn.: University of Pennsylvania Press.

Frankel, E. (1976), "Corpuscular optics and the wave theory of light: The science and politics of a revolution in physics", *Social Studies of Science,* vol. 6, pp. 141-84.

Gaston, J. (1973), *Originality and Competition in Science.* Chicago and London: University of Chicago Press.

Gaston, J. (1978), *the Reward System in British and American Science.* New York: Wiley.

Gaston, J. (ed.) (1978a), *The Sociology of Science.* San Francisco: Jossey-Bass.

Habermas, J. (1972), *Knowledge and Human Interests.* London: Heinemann.

Hanson, N.R. (1965), *Patterns of Discovery.* Cambridge: Cambridge University Press.

Hempel, C.G. (1965), *Aspects of Scientific Explanation.* New York and London: The Free Press.

Hesse, M.B. (1974), *The Structure of Scientific Inference.* London: Macmillan.

Hesse, M.B. (1980), *Revolutions and Reconstructions in the Philosophy of Science.* Brighton: Harvester Press.

Hessen, B. (1931), "The social and economic roots of Newton's Principia", in N.I. Bukharin et al. (eds.) *Science at the Crossroads,* 2nd edition (1971). London: Frank Cass.

Jacob, J.R. (1977), *Robert Boyle and the English Revolution.* Brighton: Harvester Press.

Kuhn, T.S. (1957), *The Copernican Revolution.* Harvard University Press.

Kuhn, T.S. (1959), "The essential tension", in C.W. Taylor (ed.)

Third University of Utah Conference on the Identification of Scientific Talent. University of Utah Press (reprinted in Kuhn, 1977).

Kuhn, T.S. (1961), "The function of measurement in modern physical science", *Isis,* vol. 52, pp. 161-90 (reprinted in Kuhn, 1977.

Kuhn, T.S. (1962), "The historical structure of scientific discovery", *Science,* vol. 136, pp. 760-4 (reprinted in Kuhn, 1977).

Kuhn, T.S. (1963), "The function of dogma in scientific research", in A.C. Crombie (ed.) *Scientific Change*. London: Heinemann, pp. 347-69.

Kuhn, T.S. (1964), "A function for thought experiments", in *Melanges Alexandre Koyre*. Paris: Hermann, vol. 2, pp. 307-34 (reprinted in Kuhn, 1977).

Kuhn, T.S. (1968), "The history of science", *International Encyclopedia of the Social Sciences,* vol. 14, pp. 74-83 (reprinted in Kuhn, 1977).

Kuhn, T.S. (1969), "Comment on the relations of science and art", *Comparative Studies in Society and History,* vol. 11, pp. 403-12 (reprinted in Kuhn, 1977).

Kuhn, T.S. (1970), *The Structure of Scientific Revolutions,* 2nd edn. University of Chicago Press (first published in 1962).

Kuhn, T.S. (1971), "History and the history of science", *Daedalus,* vol. 100, pp. 271-304 (reprinted in Kuhn, 1977).

Kuhn, T.S. (1971a), "Scientific growth: Reflections on Ben-David's 'scientific role'", *Minerva,* vol. 10, pp. 166-78.

Kuhn, T.S. (1974), "Second thoughts on paradigms", in F. Suppe (ed.) *The Structure of Scientific Theories*. Illinois University Press, pp. 459-82 (reprinted in Kuhn, 1977).

Kuhn, T.S. (1975), "A formalism for scientific change", in R.E. Butts and K.J.J. Hintikka (eds.) *Proceedings of the Fifth International Congress of Logic, Methodology and Philosophy of Science*. Reidel, 1977.

Kuhn, T.S. (1977), *The Essential Tension*. University of Chicago Press.

Kuhn, T.S. (1978), *Black Body Theory and the Quantum Discontinuity, 1894-1912*. Oxford: Clarendon Press.

Lakatos, I. (1970), "Falsification and the methodology of scientific research programmes", in Lakatos and Musgrave (eds.) *Criticisms and the Growth of Knowledge.*

Lakatos, I. (1971), "History of science and its rational reconstructions", in Buck and Cohen (eds.) *Boston Studies,* vol. 8. Reidel, Dordrecht.

Lakatos, I. (1976), *Proofs and Refutations.* Cambridge: Cambridge University Press.

Lakatos, I. and Musgrave, A. (eds.) (1970), *Criticism and the Growth of Knowledge.* Cambridge University Press.

Lakatos, I. (1971), "History of science and its rational reconstructions", in Buck and Cohen (eds.) *Boston Studies,* vol. 8. Reidel, Dordrecht.

Lakatos, I. (1976), *Proofs and Refutations.* Cambridge: Cambridge University Press.

Latour, B. and S. Woolgar (1979), *Laboratory Life: The Social Construction of Scientific Facts.* London: Sage.

Laudan, L. (1977), *Progress and Its Problems.* Berkeley: University of California Press.

Laudan, L. (1981), *Science and Hypothesis.* Dordrecht: Reidel.

Laudan, L. (1981a), "A confutation of convergent realism", *Philosophy of Science,* 1981.

Law, J. (1975), "Is epistemology redundant? A sociological view", *Philosophy of Social Science,* vol. 5, pp. 317-37.

Law, J. (1976), "Theories and methods in the sociology of science: An interpretative approach", in G. Lemaine et al. (eds.) *Perspectives on the Emergence of Scientific Disciplines.* Paris: Mouton, and Chicago: Aldine.

Law, J. and French, D. (1974), "Normative and interpretative sociologies of science", *Sociological Review,* vol. 22, pp. 581-95.

Lukes, S. (1974), "Relativism: cognitive and moral", *Proceedings of the Aristotelian Society,* Supplementary vol. 48, pp. 165-89.

Lukes, S. and Hollis, M. (eds.) (1982), *Rationality and Relativism.* Oxford: Blackwell.

MacKenzie, D. (1978), "Statistical theory and social interests: A case study", *Social Studies of Science,* vol. 8, pp. 35-83.

MacKenzie, D. (1979), "Karl Pearson and the professional middle class", *Annals of Science,* vol. 36, pp. 125-43.

MacKenzie, D. (1981), *Statistics in Britain, 1865-1930: The Social Construction of Scientific Knowledge.* Edinburgh:

Edinburgh University Press.

MacKenzie, D. and Barnes, S.B. (1975), "Biometrician versus Mendelian: A controversy and its explanation", *Kolner Zeitschrift fur Soziologie,* Special Edition No. 18, pp. 165-96.

MacKenzie, D. and Barnes, S.B. (1979), "Scientific Judgement: the Biometry-Mendelism controversy", in Barnes and Shapin (eds.) *Natural Order.*

Mannheim, K. (1936), *Ideology and Utopia.* New York: Harcourt, Brace & World.

Mannheim, K. (1952), *Essays on the Sociology of Knowledge.* London: Routledge and Kegan Paul.

Marx, K. (1904), *A Contribution to the Critique of Political Economy.* Chicago: C.H. Kerr.

Marx, K. and F. Engels (1965), *Manifesto of the Communist Party.* Moscow: Progress Publishers.

Marx, K. (1974), *Economic and Philosophical Manuscripts of 1844.* Moscow: Progress Publishers.

Merton, R.K. (1957), *Social Theory and Social Structure* (revised and enlarged ed.). New York: The Free Press.

Merton, R.K. (1970), *Science, Technology and Society in Seventeenth-Century England.* New York: Harper & Row (originally published in 1938).

Merton, R.K. (1973), *The Sociology of Science.* Chicago and London: University of Chicago Press (ed. with an introduction by N.W. Storer).

Merton, R.K. (1975), "Structural analysis in sociology", in P. Blau (ed.) *Approaches to the Study of Social Structure.* New York: The Free Press, pp. 21-52.

Mitroff, I.I. (1974), *The Subjective Side of Science.* Amsterdam: Elsevier.

Mulkay, M.J. (1969), "Some aspects of cultural growth in the natural sciences", *Social Research,* vol. 36, pp. 22-52.

Mulkay, M.J. (1972), *The Social Process of Innovation.* London: Macmillan.

Mulkay, M.J. (1974), "Methodology in the sociology of science", *Social Science Information,* vol. 13, pp. 107-19.

Mulkay, M.J. (1976), "Norms and ideology in science", *Social Science Information,* vol. 15, pp. 637-56.

Mulkay, M.J. (1976a), "The mediating role of the scientific elite", *Social Studies of Science,* vol. 6, pp. 445-70.

Mulkay, M.J. (1977), "Connections between the quantitative history of science, the social history of science and the sociology of science:, *Proceedings of the International Seminar on Science Studies*, Helsinki, Academy of Finland, pp. 54-76.

Mulkay, M.J. (1977a), "Sociology of the scientific research community", in I. Spiegel-Rosing and D.J. de Solla Price (eds.) *Science, Technology and Society*. London: Sage, pp. 93-148.

Mulkay, M.J. (1979), *Science and the Sociology of Knowledge*. London: Allen and Unwin.

Mulkay, M.J., Gilbert, G.N. and Woolgar, S. (1975), "Problem areas and research networks in science", *Sociology*, vol. 9, pp. 187-203.

Mulkay, M.J. and Williams, A.T. (1971), "A sociological study of a physics department", *British Journal of Sociology*, vol. 22, pp. 68-82.

Nagel, E. (1961), *The Structure of Science*. London: Routledge and Kegan Paul.

Newton-Smith, W. (1981) *The Rationality of Science*. Oxford: Routledge and Kegan Paul.

Nickles, T. (ed.) (1980), *Scientific Discovery* (2 vols.). Dordrecht: Reidel.

Phillips, D.L. (1977), *Wittgenstein and Scientific Knowledge*. London: Macmillan.

Pickering, A. (1981), "The role of interests in high-energy physics: The choice between charm and colour", *Sociology of the Sciences Yearbook*, vol. 5, pp. 107-38.

Pickering, A. (forthcoming), "The hunting of the Quark", Isis.

Polanyi, M. (1958), *Personal Knowledge*. London: Routledge and Kegan Paul.

Popper, K.R. (1959), *The Logic of Scientific Discovery*. New York: Harper & Row.

Popper, K.R. (1960), *the Poverty of Historicism*. London: Routledge and Kegan Paul.

Popper, K.R. (1963), *Conjectures and Refutations*. London: Routledge and Kegan Paul.

Popper, K. (1972), *Objective Knowledge*. London: Oxford.

Price, D.J. de Solla (1963), *Big Science, Little Science*. New York: Columbia University Press.

Putnam, H. (1975), *Collected Papers,* vol. 2. Cambridge University Press.

Quine, W.V.O. (1953), "Two dogmas of empiricism", in *From a Logical Point of View*. Harvard University Press.

Quine, W.V.O. (1960), *Word and Object*. Boston: Harvard University Press.

Ravetz, J.R. (1971), *Scientific Knowledge and Its Social Problems*. Oxford: Clarendon Press.

Roll-Hansen, Nils (1979), "Experimental method and spontaneous generation: the controversy between Pasteur and Pouchet, 1859-64. *Journal of the History of Medicine,* 34, pp. 273-292.

Rose, H. and Rose, S. (1976) (eds.), *The Political Economy of Science*. London: Macmillan.

Rudwick, M.J.S. (1972), *The Meaning of Fossils*. London: Macdonald.

Rudwick, M.J.S. (1974), "Darwin and Glen Roy: A 'great failure' in scientific method?", *Studies in the History and Philosophy of Science,* vol. 5, no. 2, pp. 97-185.

Rudwick, M.J. (1976), "The emergence of a visual language for geological science", *History of Sciencel,* vol 14, pp. 149-95.

Ruse, M. (1979), *The Darwinian Revolution*. Chicago: University of Chicago Press.

Ruse, M. (1981), *Is Science Sexist?* Dordrecht: Reidel.

Sandow, A. (1938), "Social factors in the origin of Darwinism", *The Quarterly Review of Biology,* vol. 13, pp. 315-26.

Scheffler, I. (1963), *The Anatomy of Inquiry*. New York: Knopf.

Schefller, I. (1967), *Science and Subjectivity*. New York: Bobbs-Merrill.

Shapin, S. (1975), "Phrenological knowledge and the social structure of early nineteenth-century Edinburgh", *Annals of Science,* xxxii, pp. 219-43.

Shapin, S. (1979), Homo phrenologicus: Anthropological perspectives on an historical problem", in Shapin and Barnes (eds.) *Natural Order,* pp. 41-71.

Shapin, S. (1979a), "The politics of observation: Cerebral anatomy and social interests in the Edinburgh phrenology disputes", in Roy Wallis (ed.) *On the Margins of Science: The Social Construction of Rejected Knowledge,* Sociological Review Monographs, vol. xxvii (Keele, Staffs), pp. 139-78.

Shapin, S. (1980), "A view of scientific thought" [Review of Fleck, *Genesis and Development of a Scientific Fact*], Science, (7 March 1980), pp. 1065-66.

Shapin, S. (1980a), "A course in the social history of science", *Social Studies of Science*, x, pp. 231-58.

Shapin, S. (1980b), Untitled review of Geison, Michael Foster and the Cambridge School of Physiology, *Isis*, pp. 146-49.

Shapin, S. (1980c), "Social uses of science", in G.S. Rousseau and Roy Porter (eds.) *The Ferment of Knowledge: Studies in the Historiography of Eighteenth-Century Science*. Cambridge: Cambridge University Press, pp. 93-139.

Shapin, S. (1981), "Of gods and kings: Natural philosophy and politics in the Leibniz-Clarke disputes", *Isis*, lxxii, pp. 187-215.

Shapin, S. (1981a), "Licking Leibniz" [Review of Hall, Philosophers At War: The Quarrel Between Newton and Leibniz], *History of Science*, xix, pp. 293-305.

Shapin, S. (1982), "History of science and its sociological reconstructions", *History of Science*, xx, pp. 157-211.

Shapin, S. (1982a), "'Nibbling at the teats of science': Edinburgh and the diffusion of science in the 1830s", in Ian Inkster and J.B. Morrell (eds.) *Metropolis and Province: The Social History of British Science 1780-1850*. London: Hutchinson; Philadelphia: University of Pennsylvania Press.

Shapin, S. (forthcoming), "Making out Newton: On the interpretation of scientific texts".

Shapin, S. and Barnes, B. (1976), "Head and hand: Rhetorical resources in British pedagogical writing, 1770-1850", *Oxford Review of Education*, vol. 2 no. 3, pp. 231-54.

Shapin, S. and Barnes, B. (1977), "Science, nature and control: Interpreting mechanics' institutes", *Social Studies of Science*, vol. 7 no. 1, pp. 31-74.

Shapin, S. and Barnes, B. (1977a), "Where is the edge of objectivity?" Douglas, *Implicit Meanings, British Journal for the History of Science*, x, pp. 61-66.

Shapin, S. and Barnes, B. (1979), "Darwin and social Darwinism: Purity and History", in *Natural Order* (see above), pp. 125-42.

Shapin, S. and Schaffer, S. (forthcoming), Leviathan and the Air Pumps: Hobbes, Boyle and the Politics of Experiment.

Sklair, L. (1973), *Organized Knowledge*. London: Hart-Davis, MacGibbon.

Stark, W. (1958), *The Sociology of Knowledge*. London: Routledge and Kegan Paul.

Storer, N.W. (1966), *The Social System of Science*. New York: Holt, Rinehart and Winston.

Teich, M. and Young, R.M. (eds.) (1973), *Changing Perspectives in the History of Science*. London: Heinemann.

Thackray, A. (1974), "Natural knowledge in cultural context", *American History Review*, vol. 74, pp. 672-709.

Tobey, R.C. (1971), *The American Ideology of National Science*. Pittsburgh, Pa.: University of Pittsburgh Press.

Turner, R.S. (1971), "The growth of professorial research in Prussia, 1818 to 1848 - Causes and context", in R. McCormmach (ed.) *Historical Studies in the Physical Sciences*, vol. 3, pp. 137-82.

Winch, P. (1958), *The Idea of a Social Science*. London: Routledge and Kegan Paul.

Winch, P. (1964), "Understanding a primitive society", *American Philosophical Quarterly*, vol. 1, pp. 307-24.

Winsor, M.P. (1976), *Starfish, Jellyfish, and the Order of Life*. Yale University Press.

Wittgenstein, L. (1953), *Philosophical Investigations*. Oxford: Blackwell.

Wittgenstein, L. (1964), *Remarks on the Foundations of Mathematics*. Oxford: Blackwell.

Woolgar, S.W. (1976), "Writing on intellectual history of scientific development: The use of discovery accounts", *Social Studies of Science*, vol. 6, pp. 395-422.

Woolgar, S.W. (1976a), "The identification and definition of scientific collectivities", in G. Lemaine et al. (eds.) *Perspectives on the Emergence of Scientific Disciplines*. Paris: Mouton and Chicago: Aldine.

Woolgar, W. and Latour, B. (1979), *Laboratory Life: The Social Construction of Scientific Facts*. London: Sage.

Wilson, B. (ed.) (1971), *Rationality*. Oxford: Blackwell.

Wynne, B. (1976), "C.G. Barkla and the J phenomenon", *Social Studies of Science*, vol. 6, nos 3 and 4, pp. 307-47.

Wynne, B. (1979), "Physics and psychics: Science, symbolic action and social control in late Victorian England", in Barnes and Shapin (eds.) *Natural Order* (1979).

Young, R.M. (1969), "Malthus and the evolutionists", *Past and Present,* vol. 43, pp. 109-45.

Young, R.M. (1973), "The historiographic and ideological contexts of the nineteenth-century debate on man's place in nature", in Teich and Young (eds.) *Changing Perspectives in the History of Science* (1973).

Larry Laudan

THE PSEUDO-SCIENCE OF SCIENCE?*

"To ask questions of the sort which philosophers address to themselves is usually to paralyse the mind...." David Bloor[1]

1. Introduction
2. The Attack on 'Philosophical' Views of Belief
3. The 'Scientific' Character of the Strong programme
4. The True, the Rational and the Successful
    4.1 Epistemic Symmetry
    4.2 Rational Symmetry
    4.3 Pragmatic Symmetry
5. The Avowed Primacy of the 'Sociological Turn'
6. Conclusion

1. Introduction

After several decades of benign neglect, the *content* of science has once again come under the scrutinous gaze of the sociologist of knowledge. Aberrant Marxists, structuralists, Habermasians, 'archeologists of knowledge' and a host of others have begun to argue (or, sometimes, to presume largely without argument) that we can give a *sociological* account of why scientists adopt virtually *all* of the specific beliefs about the world which they do. More than this, it is often claimed that *only* via sociology (or its cognates, anthropology and archaeology) can we hope to acquire a 'scientific' understanding of science itself. The older sociological tradition, which tended to take a hands-off policy where 'sound' scientific belief was concerned, has been variously indicted by the new wave as lacking the courage of its convictions, treating science as

J. R. Brown (ed.), Scientific Rationality: The Sociological Turn, 41–73.

'sacred' and unimaginatively selling short the explanatory resources of a robust sociology of knowledge.

Within the English-speaking community, this point of view has received its most forceful and its most frequently-cited formulation in what has come to be known as 'the strong programme in the sociology of knowledge'. During the last few years, the approach denoted by that phase has received expression at the hands of, or high praise from, a number of historians, philosophers and sociologists of science, including Hesse, Rudwick, Caneva, Barnes, Bloor, Shapin and numerous others.[2] The ability of the strong programme to attract this growing number of adherents is unquestionably impressive. But one is as perplexed as impressed when one examines some of the staggering tensions between alternative formulations of the strong programme. As Manier[3] among others has pointed out, the two chief authors of the strong programme -- Barnes and Bloor -- say very different things about what that programme amounts to. Mary Hesse, a sympathetic commentator on their work, has recently offered her own version of the strong programme which -- as she herself stresses -- is frequently at odds with both Bloor and Barnes. Obviously, if one is free to make the strong programme into whatever one wants it to be, we have a ready account of its broad appeal, but little else. If we would understand what the strong programme amounts to, I suggest that a good beginning would involve us in a discussion of the book which is generally regarded as the first and still probably the most articulate formulation of that programme, namely David Bloor's *Knowledge and Social Imagery*. In this paper, I shall try to piece together what Bloor's version of the strong programme amounts to, and examine what it presupposes. How far the analysis offered here applies to other proponents of the strong programme will, except for occasional asides, be left to the reader's imagination.

The first crucial thing to note about the strong programme is that it is *not* a sociological theory, in any customary sense of that term. It specifies no detailed causal or functional mechanisms and no laws. It is, rather, a *meta-sociological* manifesto. It lays down certain very general characteristics which any adequate sociology of knowledge should possess. It is 'programmatic' in the strict sense; it must be approached, and I think was intended to be approached, as a set of regulative principles about what sort of theories sociologists should aspire to. Its four constituent 'theses' are designed as constraints on the theories which are admissible into sociology. It is important to understand this about the character of the strong programme, since one evaluates regulative principles differently than one evaluates specific theories about social structure and social process. programmatic pronouncements, and here the strong programme is no exception, are generally too amorphous to be put directly to empirical test. They are judged rather by what we may call their *plausibility*. We ask: Is it reasonable to adopt such constraints? Are there any arguments for preferring the proposed demands rather than other conflicting requirements?

Knowledge and Social Imagery is simultaneously a sustained tirade against philosophers of almost all persuasions and a lengthy articulation of a 'new' and ambitious programme for the sociology of knowledge. The two themes are not unconnected. As Bloor sees it, philosophers (under the guises of 'epistemology' and 'philosophy of science') have attempted to monopolize the cognitive study of knowledge, especially scientific knowledge, leaving only the fringes and dregs -- the irrational residuum -- to psychologists and sociologists.[4] Hoping to beat philosophers at their own territorial game, Bloor sets out to re-define the disciplinary boundaries for the study of science, giving sociology pride of place, leaving a limited scope for psychology, and dealing philosophers, on the strength of their prior track record, largely out of the new game altogether. Bloor's 'strong programme in the sociology of knowledge' is ostensibly a set of principles about how sociologists should approach the problem of explaining scientific belief. Beyond that, it entails that (what Bloor takes to be) the dominant philosophical approach to knowledge is hopelessly unscientific, non-naturalistic and unempirical; to make matters worse, philosophers have become proponents of a 'mystical' view of knowledge, the chief function of which is to preserve the 'sacred' character of science in the face of sociologists who might profane it.

Whenever philosophers and sociologists of knowledge discuss such matters, there is scope for a great deal of smugness and self-righteousness on both sides. Quite apart from the rival disciplinary interests which are at stake, both parties can generally point to a range of important problems their opponents are ignoring, issues they are over-simplifying, controversial presuppositions they are making, and the like. Bloor himself (as I suspect he would concede) is no more averse to playing this game than the rest of us; indeed its playing out is one of the dominant narrative threads of *Knowledge and Social Imagery*. As a philosopher responding to Bloor's work, I have found myself tending to play the game, too; noting to myself, for instance, the places where Bloor has either stepped on philosophical toes or offended philosophical sensibilities. Elements of that attitude are doubtless still present in this much expurgated version of this essay!

What I have tried to do in the following however, is to focus chiefly on the structure of Bloor's arguments, without engaging in any special pleading for the disciplinary interests of philosophy. Insofar as the considerations adduced here have any force, they are addressed to hard-headed sociologists and philosophers alike, and to anyone else who accepts that one's beliefs should be shaped (insofar as they can be) by the strength of the evidence and arguments which can be adduced for them.

The strategy of Bloor's book will dictate the character of this response. Bloor begins by assaulting what he regards as the two philosophical alternatives to his own position. He labels them 'teleology' and 'empiricism' respectively. Having disposed of these alternatives, he then calls for a 'scientific' approach to the study of knowledge, and sketches out a number of demands which any genuinely scientific approach must satisfy. These demands constitute 'the strong programme in the sociology of knowledge'. There follows an account of why ideas like those embodied in the strong programme have been resisted through recent Western intellectual history. The remainder of the book is devoted to specific case studies -- chiefly in the sociology of mathematics -- which are meant to illustrate how a sociologist committed to the strong programme might seek to implement and particularize it.

I shall follow the chronology of Bloor's argumentative strategy. Specifically, I shall suggest that:

i). Bloor has selectively ignored much of the relevant philosophical literature and has saddled 'philosophers' with views that few (if any) would espouse;

ii). Bloor has *not* established that *any* elements of the strong programme are more 'scientific' than their contraries;

iii). Quite apart from their scientific status, the theses of the strong programme are of very different sorts. Some are so unproblematic as to be almost gratuitous. Others (specifically the 'thesis of symmetry') are not made plausible by any of Bloor's arguments and, if construed literally, would fundamentally undermine existing explanatory mechanisms in both philosophy and the social sciences. Limitations of space and competence preclude any lengthy discussion of Bloor's fascinating examples from the history and sociology of mathematics; but that is no serious loss, since it is not Bloor's sociology I want to take exception to but rather his meta-sociology which I find unconvincing.

2. The Attack on 'Philosophical' Views of Belief

Bloor's central concern is with the explanation of scientific belief, especially of what one might call 'theoretical belief' (i.e., a belief in the theories and theoretical entities which populate the conceptual universes of natural scientists.)[5] Why is it that scientists came to hold the theoretical beliefs that they do? This is, of course, one of the classical questions of epistemology and philosophy of science. As Bloor sees it, however, the philosopher makes a wrong turn at the outset. Before he seeks to explain a belief, the philosopher attempts to determine its truth status and its rationality status (i.e., whether the belief is reasonable or rational). Depending upon the answer to those evaluative inquiries, the philosopher will adopt radically different approaches to explaining the belief. If the philosopher is a 'teleologist' (and Bloor evidently thinks most are), he will insist that no explanation whatever can be given for true or rational beliefs. The 'teleologist' insists that such beliefs just happen, that they are literally uncaused. it is only false or irrational beliefs for which an explanation can appropriately be sought. As Bloor summarizes the 'teleological' position:

> The general structure of these explanations stands out clearly. They all divide behavior or belief into two types: right and wrong, true or false, rational or

> irrational. They then invoke causes to explain the negative side of the division. Causes explain error, limitation and deviation. The positive side of the evaluative divide is quite different.... Here causes do not need to be involved.[6]

The so-called 'empiricist', by contrast, believes that explanations can be given for both true and false beliefs, as well as for rational and irrational ones. But he insists that true beliefs have *different* sorts of causes than false ones and that rational beliefs are produced differently from irrational ones.

What both teleologists and empiricists share in common is a conviction that an appraisal of the epistemic status and the rationality status of a belief is relevant to the mechanisms (if any) which we subsequently invoke to explain that belief. As we shall see, it will be Bloor's claim that this manner of proceeding is intrinsically 'unscientific'. Indeed, Bloor will go on to object not only to these prior appraisals, but to *any* use of the epistemic or rationality status of a belief in its explanation. Although Bloor has this crucial bone to pick with both teleologists and empiricists, he is particularly scornful of the teleologist for the latter's insistence that true and rational beliefs are uncaused. This teleological model violates another of Bloor's 'scientific' theses: specifically, the claim that all beliefs are caused.

If Bloor's caricatures were to be accepted, we should believe that most philosophers (being 'teleologists') have maintaianed that there is literally nothing that causes us to believe what is true and that nothing is causally responsible for rational action and rational belief. But Bloor's analysis of the philosophical tradition will not stand up to scrutiny. For as long as we know anything about the history of philosophy, epistemologists have been concerned to explain how to discover the true and the rational. The suggestion that most philosophers have believed that true beliefs just happen, that rational behavior is uncaused, that only 'aberrant' belief is part of the world's causal nexus, is hard to take seriously.

It is true that many philosophers have suggested that true or rational beliefs are not to be attributed to *sociological* causes. But unless we are to imagine that sociology has a monopoly on causes, the denial that true or false beliefs have social causes is manifestly not equivalent to the assertion that true and rational beliefs are *uncaused*!

What I am saying, to use Bloor's language, is that few if any major philosophers have been 'teleologists.' More specifically, I am not aware of *any* philosopher of science who has ever asserted that true or rational scientific beliefs have no causes. Bloor's lengthy and cogently argued attack on this position is very largely a matter of flogging a mythical horse. Most philosophers, every bit as much as their sociological counterparts, are committed to the view that beliefs, *whatever their epistemic status*, are part of the world's causal network.

So, insofar as any large group of philosophers fits Bloor's stereotypes, it is the 'empiricists' who constitute the real opposition. But even here, there are crucial distinctions to be drawn before we can be sure who Bloor's targets are; for Bloor lumps together issues that need to be carefully distinguished. In particular, he tends to treat issues of epistemology and issues of rationality as if they were one and the same. This slide occurs both in Bloor's attack on the philosophers, and (as we shall see later) in the development of his own position. For instance, he characterizes the 'empiricist' philosophers as holding *both* that true beliefs are to be explained differently from false ones *and* that rational beliefs are to be explained differently from irrational beliefs. There is *no* logical connection between these two theses; more to the point, thinkers who hold the one often deny the other. Thus, Imre Lakatos, whom Bloor singles out for criticism on these matters, does *not* maintain that true beliefs are to be explained differently from false ones, only that rational and irrational beliefs are to be explained differently. Again, some psychologists maintain that certain true and false beliefs are to be differently explained (that is why they distinguish between 'veridical' perceptions and hallucinations for instance), yet they need draw no such distinction between rational and irrational beliefs. Economists explain rational and irrational economic behavior by different models yet they are not committed to different explanatory programmes for true and false beliefs. By assimilating these very different ways of evaluating beliefs, Bloor has persuaded himself that arguments against the viability of one inevitably cut against the viability of the other. As we shall see below, this becomes the source of much mischief in the formulation of the strong programme.

For now it is sufficient to note the following:

i). few if any philosophers of science are 'teleologists' in Bloor's sense. He is fighting against straw men in combatting the view that beliefs literally have no causes.

ii). of the philosophers who subscribe to the view that the truth or rationality of a belief is germane to its explanation, very few insist on the *causal* relevance of both epistemic and rational considerations. Bloor's broadsides against the 'empiricists' misfire insofar as they rest on a neat parallel between those doctrines.

To this point, I have tried to suggest that Bloor has done less than justice to the philosophical tradition of explaining belief. His opponents are not quite as crude as he sometimes depicts them. But Bloor still has an ace up his sleeve, for he will charge over and again that the philosophical approach to these issues -- whether represented by the 'teleologists' or the 'empiricists' -- takes an inexcusably unscientific approach and that Bloor's own 'strong programme' represents the only appropriate 'scientific' counterpoint to the philosopher's *modus operandi*. It is to that claim that we should turn.

3. The 'Scientific' Character of the Strong Programme

The fourth section of this paper will discuss the various theses of the strong programme in some detail. For now, I am concerned not so much with their precise explication but rather with Bloor's claim that they represent a robustly 'scientific' approach to the problem of human belief. At the outset, I can do no better than quote Bloor's own characterization of the sociology of knowledge as the strong programme would require it to look:

1. It would be causal, that is, concerned with the conditions which would bring about belief or states of knowledge. Naturally there will be other types of causes apart from social ones which will cooperate in bringing about belief.
2. It would be impartial with respect to truth and falsity, rationality or irrationality, success or failure. Both sides

of these dichtomies will require explanation.

3. It would be symmetrical in its style of explanation. The same types of cause would explain, say, true and false beliefs.
4. It would be reflexive. In principle its patterns of explanation would have to be applicable to sociology itself....[6]

Bloor justifies these theses by asserting that they represent the principles of inquiry to which every genuine scientist, natural or social, is committed.[7] These principles "embody the same values which are taken for granted in other scientific disciplines".[8] The strong programme "possess a certain kind of moral neutrality, namely the same kind as we have learned to associate with all the other sciences."[9] Bloor insists that to deny these theses -- especially theses (2) and (3) -- in favour of certain rival positions "would be a betrayal... of the approach of empirical science".[10] He even suggests that if a robust sociology, such as one based on theses 1 through 4, "could not be applied in a thorough-going way to scientific knowledge it would mean that science could not scientifically know itself."[11] "The search for laws and theories in the sociology of science," says Bloor, "is absolutely identical in its procedure with that of any other science."[12] He urges us to emulate those "whose confidence in science and its methods is wellnigh total -- those who utterly take it for granted."[13] He is "more than happy to see sociology resting on the same formulations and assumptions as other sciences."[14] Bloor insists in the conclusion to *Knowledge and Social Imagery* that he has arrived at his view of knowledge by acting on a straightforward strategy: "Only proceed as the other sciences proceed and all will be well."[15] As these passages make clear, a central motivation and justification for the strong programme is that, unlike approaches to belief in the philosophy of science and the theory of rationality, it embodies a genuinely scientific attitude towards science.

Accordingly, an obvious question to concern ourselves with is simply this: do the other sciences, which Bloor takes as his paradigm, have features which would justify his claim that no account of belief can be 'scientific' which violates theses (1) to

(4)? Or more likely, is there any good reason to believe that (1) to (4) are warranted by prevailing scientific *praxis*?

I must confess to a great deal of uneasiness in asking such questions for, unlike Bloor, I am not confident that what we call 'the sciences' have any special set of methodological principles or epistemic credentials that clearly sets them off from other supposedly 'non-scientific' forms of cognition. What I am confident about is the claim that no one, philosopher or sociologist, has yet set out any acceptable account of what cognitive or methodological features demarcate the sciences from the non-sciences. At a time when many philosophers despair even of the *possibility* of drawing a neat distinction between the scientific and the non-scientific, Bloor believes not only that there is a clear demarcation to be found, but that he is already in possession of it. He must believe that, for otherwise he would have no ground whatever for claiming his approach to be distinctly scientific. I am not so concerned as Bloor is about whether the sociology of science is 'scientific'. I would like it to be interesting, which it sometimes is, and well-tested which it usually is not; beyond that, my demands are fairly modest. But Bloor is keen to 'scientize' sociology; it is the 'scientificity' of the strong programme which is extolled as its greatest virtue. Has he pulled it off? Well, what is scientific about theses (1) to (4)?

Consider the thesis of causality. Is all scientific knowledge causal? Hardly. Whether one looks to portions of quantum mechanics, statistical mechanics or classical kinematics, one looks in vain for an overriding commitment to causal processes. Equally, there are striking froms of non-scientific knowledge (e.g. metaphysics and theology) which are avowedly causal. Neo-Kantians may take comfort from Bloor's claim that all scientific knowledge is causal but scientists and philosophers of other persuasions are not apt to accept without further ado the claim that causal talk is either a necessary or a sufficient condition for being 'scientific'. Let me stress that I share Bloor's view that we would like to be able to identify the causes of belief; but I would urge that our managing to do so would no more guarantee the 'scientific' character of our enterprise than our failure to do so would necessarily make us unscientific. Properly construed, about which more anon, thesis (1) of the strong programme is innocuous; it also has nothing whatever to do with whether sociology is scientific. More to the point, one can accept that beliefs are caused without accepting that such causes are invariably or even generally sociological in character.

Thesis (2), the tenet of 'impartiality', occurs in no science of which I am aware. Moreover, it is largely gratuitous. Indeed, given Bloor's first thesis to the effect that one should seek the causes of all scientific beliefs, his second thesis is redundant, for it is a corollary of the first. This obviously is no objection to it, and Bloor nowhere suggests that his theses are logically independent. But insofar as the thesis of impartiality is parasitic on the thesis of causality, its 'scientific' status is as open to challenge as the first. Thesis (4), the thesis of reflexivity, is equally redundant since it too follows from (1), although why Bloor thinks it is espcially scientific is opaque to me. If we are going to have a perfectly *general* theory about how beliefs are formed, then such a theory (if it is an object of belief) will necessarily be self-referential, and this will be the case whether the theory is scientific or not. Indeed, an avowedly *non-scientific* theory of belief formation would, to be consistent, have to be similarly self-referential. Three of the four theses of the strong programme are thus relatively uncontroversial, but their scientific status has yet to be established.

Thesis (3), the symmetry thesis, is in a class of its own. Where my argument about causality, reflexivity and impartiality was to the effect that Bloor has not yet established a case that they are 'scientific', I want to suggest that the symmetry principle seems to run *against* the best established precedents in the natural sciences. To a first aproximation (and this will be discussed in detail below), the symmetry principle demands that the same causal mechanisms are to be invoked for explaining all cases of belief, whether those beliefs are true, false, rational, irrational, successful or unsuccessful. Bloor suggests that the methodological principles of simplicity and economy require such symmetry. But in what science is it assumed that all events are to be explained by the same sorts of causal mechanisms? Physicists do not explain gravitational and electrical phenomena by invoking the same causal processes. Chemists do not explain bonding and osmosis by utilizing similar models. Geologists, to a man, would resist the suggestion that erosion and uplift might be the result of similarly acting causes.

When scientists discover *prima facie* differences in the involvement of causal agents in the world, they do not hesitate to use different models and mechanisms for explaining them. Especially when it comes to the difference between rational and irrational beliefs, many philosophers and social scientists believe themselves to be dealing with just such a distinction in the causal order. They may be wrong in that belief, of course; it may turn out that all beliefs are produced in the same way. But it is scarcely 'scientific' to attempt to settle that contingent question by stipulative fiat. The symmetry thesis could lay claim to scientific status only *after* it had been established that the generation of belief was causally homogeneous. In the absence of independent evidence for that homogeneity (evidence which Bloor does *not* produce), the symmetry thesis has the effect of resolving an empirical question by *a priori* means.

Further problems abound: If we cannot say what cognitive features are unique to the sciences, how can we possibly attempt to emulate them? Even if we did know what made science 'scientific', by what license could we "utterly take for granted" the legitimacy of its methods? How can Bloor, who protests against the assimilation of science to the sacred, be consistent in urging that the only legitimate mode in which to study science is the scientific mode? I have not found ready answers to any of these questions in Bloor's text. Bloor himself concedes that "the charge of scientism against his work is well aimed."[16] But notwithstanding that commendable candor, one is hard pressed to resist the conclusion that -- where the scientific status of the strong programme is concerned -- Bloor aimed for more than he has delivered.

Quite apart from Bloor's specific failure to make the case that the strong programme is 'scientific', the general enterprise seems to put the cart before the horse. If one is to take a genuinely empirical approach to the study of scientific knowledge, one is well-advised to leave the question of exactly what characterizes scientific knowledge open until the relevant data are in hand. If Bloor already knows what metholdological procedures and regulative principles characterize 'science', then what is the object of the exercise? One might have imagined, more modestly, that one would postpone any decision about whether the strong programme was scientific until after one had carefully studied -- sociologically or otherwise -- what features the knowledge systems we call 'scientific' exhibit. There is something profoundly paradoxical in saying that we are setting out scientifically to figure out what central features science has.

In this section I have limited my treatment of the strong programme to the very specific question of its 'scientific' status. There is much to be said in favor of several theses of the strong programme, as well as some challenges for it to overcome. Those issues, to be discussed below, are not germane here. What is crucial is the recognition that the pretensions of the strong programme to honorfic status as a science are thus far unfounded. The strong programme has not been shown to be any more scientific than the approaches of the 'teleolgist' or the 'empiricist', not to mention numerous other alternatives. If a strong case can be made for the theses of the strong programme, that case will have to rest on the specific arguments that can be given for each of its constituent theses; the attempt at wholesale legitimation by depicting the strong programme as the unique or preferred scientific option will not wash. Mutton dressed up as lamb remains mutton all the same!

4. The True, the Rational and the Successful

What, then of the four theses of the strong programme? The first, the thesis of causality, is relatively unproblematic. I do not thereby mean to suggest that our conceptions of doxastic and social causation are crystal clear; quite the reverse is true, of course. But it is my impression (although not Bloor's) that practically every sociologist and philosopher of science who has discussed the question of human belief has acquiesed in the claim that something or other causes us to believe what we believe. the idea that the mental states we denote by the term 'beliefs' might be outside the causal order, that there might literally be no causal antecedents to our beliefs, is as anathema to most philosophical thinking as it is to 'scientific' thinking. Even Bloor's *betes noir,* the 'teleologists', who were supposedly committed to the view that rational behavior was a part of man's nature, would have used the causal idiom to express their view. The causal thesis is perfectly sound as a regulative principle, enjoining us to seek the causes on which beliefs depend; but it is scarcely unique to the strong programme nor -- given how weak a claim it makes -- can it be said to be what renders the strong programme 'strong'.

The thesis of impartiality and reflexivity are, as already noted, virtual corollaries of the thesis of causality. It would be on pain of contradiction that one could accept the causal thesis and deny these other two. They are every bit as sound, and every bit as uncontroversial, as the first thesis. Again then, one looks in vain for some teeth in these elements of the strong programme.

But one does not have to look far, since the thesis of symmetry alone is sufficient both to save the strong programme from the charge of toothlessness and to guarantee that Bloor's approach will meet a great deal of opposition. Boldly put, the thesis of symmetry is a strong formulation of *cognitive relativism* and it will provoke all the usual counter moves to relativism. I do not want to rehearse all those familiar themes here (e.g., "Does not all communication presuppose certain common rules of inference?" and "Aren't there certain logical truths which are not culture and context bound?") Many arguments against relativism or symmetry are *transcendental* arguments (of the form "a pre-condition for the possibility of x is y..."). I share Bloor's impatience with the transcendental and would not expect him to be moved by such considerations. Nor, come to that, is it my brief to attack relativism *tout court*. There are some forms of cognitive relativism which seem to be unavoidable. But the omnibus relativism entailed by the symmetry thesis is, I think, without warrant. It is the purpose of the present section to explain why.

But before I can do that, we must try to pin down more precisely than Bloor does what the symmetry thesis amounts to. Although Bloor's formulation of the thesis quoted above refers only to the contrast between true and false beliefs, his examples and further discussion make sense only if we construe the symmetry theses as referring additionally to at least the same contrasts mentioned in the impartiality thesis -- namely, between rational and irrational beliefs and between successful and unsuccessful ones. For purposes of exposition, then, and I think this is faithful to the spirit of Bloor's enterprise, we might break the thesis of symmetry down into three constituent sub-theses:

i) *epistemic symmetry:* true and false beliefs are to be explained by the same types of cause.

ii) *rational symmetry:* rational and irrational beliefs are to be explained by the same types of cause.

iii) *pragmatic symmetry:* successful and unsuccessful beliefs are to be explained by the same types of cause.

Since these three provocative doctrines raise very different issues, and lead to very different variants of relativism, I want to deal with them separately. But before I turn to that examination, there is one preliminary problem which they all share.

At the heart of the symmetry thesis is a fundamental unclarity which makes the assessment of that thesis difficult. I refer to the notion of *sameness of type of cause.* To be told that we must explain true or false beliefs, and rational or irrational ones, by the 'same types of cause' requires -- at least informally -- the specification of a taxonomy of causal types. Until we know how to classify causes, we are helpless when confronted by the injunction to use the same sorts of causes for, say, true and false beliefs. This worry is more than pedantic since it decisively affects one's judgment of the strong programme. For instance, is the citing of an agent's reason for his belief the same kind of cause as the explanation of that belief in terms of the agent's socio-economic status? Is a discussion of the neuro-physiological determinants of a belief the same kind of cause as a psychoanalytic explanation of the same belief? If the answer to both questions is yes (perhaps on the grounds that all four explanations are 'naturalistic' in kind) then the symmetry thesis is innocuous and uncontroversial. If on the other hand, both answers are no, then we are seemingly dealing with four different sorts of causes and the symmetry principle would accordingly entail that every belief is to be explained by the same one (or the same combination) of them. On that construal, the symmetry thesis is strongly monistic, arguing that some one (or possibly a specific combination) of these causes enjoys a monopoly in the production of belief states.

Bloor, I believe, never clarifies where he stands on this issue and thus leaves his reader unclear about how to proceed. But perhaps what Bloor is driving at is this: *whatever* causal mechanisms we find useful for explaining beliefs, we should invoke them without reference to the epistemic or the rationality or the pragmatic status of the beliefs we want to explain. Bloor says as much when he writes: "All beliefs are to be explained in the same way regardless of how they are evaluated."[17] Construed sympathetically, the talk of 'same types of cause' is perhaps just a *facon de parler*. On this construal, what the symmetry thesis is really asserting is the *causal or explanatory irrelevance* of one's knowledge of the truth, rationality or success of a belief in giving a 'naturalistic' account of how an agent came to have the belief. The symmetry thesis so construed is an assertion of that irrelevance. This reading of the symmetry thesis is given added plausibility by the work of Bloor's colleague, Barry Barnes, whose version of the symmetry thesis is this:

> What matters is that we recognize the *sociological* equivalence of different knowledge claims. We will doubtless continue to evaluate beliefs differentially ourselves, but such evaluations must be recognized as having no relevance to the task of sociological explanation; as a methodological principle we must not allow our evaluation of beliefs to determine what form of sociological account we put forward to explain them.[18]

Accordingly, in what follows, I shall assume that Bloor's various symmetry theses amount to claims of causal independence of belief from epistemic, rationalistic and pragmatic considerations.[19]

4.1 Epistemic Symmetry

I think Bloor is quite right in asserting, with respect to our theoretical beliefs, that their truth status is largely if not entirely irrelevant to their explanation. This is an important insight of the strong programme and its rationale needs to be spelled out as clearly as possible. I am not sure my reasons for arguing the causal irrelevance of truth and falsity are the same as Bloor's, but it is possibly worth setting those reasons out briefly since they effectively entail the 'de-epistemologizing' of cognitive sociology.

When it comes to scientific theories, the most we can hope to 'know' about them (even in a modest sense of that term) is that they are false. Arguments familiar since Hume, and elaborated on at great length by Popper, show that we are never in a position to be reasonably confident that a theory is true. Still worse, we cannot reasonably assume even that a theory is 'approximately ture' (in any extant sense of that term).[20] If we once appreciate these facts about ourselves as knowers, it follows that we can not conceivably invoke the fact that a theory is true to explain anything whatever about its doxastic fortunes. Knowledge of a theory's truth is radically transcendent. This transcendence entails the epistemic version of the symmetry thesis since we are never in a position to partition theories into the true and false and then proceed to explain beliefs in them differently on account of their truth status.

This is not to assert (as I believe Bloor would) that *even if* we could know which theories were true and which false, it would inevitably be the case that true and false beliefs arise through the same causal mechanisms. *If* the truth status of theories were accessible to us, it would then be an *empirical* question whether true and false beliefs are differently produced. But because of that radical inaccessibility, we cannot put the issue to the test; and that self-same inaccessibility undercuts any possibility of asymmetrically explaining beliefs in true and false theories.

4.2 Rational Symmetry

As any card-carrying philosopher will tell you, rationality is a many-splendored thing. And that is too bad, because the multiple sense of 'rational' make it difficult to test the claim that rational and irrational beliefs are similarly caused. On one familiar view of rationality, for instance, being rational involves meting out one's degree of belief in accordance with the probability calculus. On another view, it involves basing one's beliefs on legitimate modes of logical inference. On still another, it comes down to adopting beliefs which conduce to one's cognitive ends. To make matters worse, 'rational' functions both as a normative and as a descriptive concept.

It would take us much too far afield to explore the ramifications of the symmetry thesis for every extant model of rationality. Nor is that necessary. If we can show that there is even one sense of rationality which is *causally* relevant to the explanation of beliefs, then we will have shown that the thesis of rational symmetry -- which insists that evaluative appraisals of a belief are never relevant to its explanation -- is ill-conceived.

I shall work with a modest notion of rational action and rational belief. It is neither a sophisticated nor a subtle view of rationality, but it will be sufficient for present purposes. On this model, a rational agent is one who has various goals and various prior beliefs about the world. His rationality consists in his engaging in a process of ratiocination in order to ascertain what course of action his goals and prior beliefs commit him to. To adopt a belief rationally, the agent must be able to specify reasons, relative to his goals and background knowledge, for adopting that belief rather than its negation. On this not unfamiliar account, a belief is rational or reasonable provided the agent can give reasons for it and can show that those reasons were antecedent to the adoption of the belief.

This is a *causal* theory of rationality, in so far as it avers that reasons can, and often do, function as the causes of belief. It rests on a contrast between beliefs that result from a process of ratiocination and reflection and those that do not. It insists that there are specific mechanisms for the generation of rational beliefs (among others, various inferential mechanisms), which are not involved in the generation of unreasoned beliefs.

Where the symmetry principle insists that rational and irrational beliefs are to be explained indifferently, the suggestion here is that rational and irrational beliefs arise in different ways and thus are to be explained by different mechanisms. It seems hard indeed to resist such a conclusion. If there are some beliefs arrived at by ratiocination and others which are not (and this is surely an uncontroversial claim), then it is evidently sound tactics to insist that our explanations of beliefs should reflect those differences.

What is being argued here is that "explanation by reasons" is a distinct species of causal explanation different from, say, familiar psychoanalytic and socio-economic modes of explanation. Some beliefs, caused by reasons, should be explained by citing the reasons. Others, caused perhaps by the direct action of social and psychological forces unmediated by reasons, call for a very different sort of explanation. Such differences raise very serious obstacles to the thesis of symmetry. More importantly, it is vital to stress that the invocation of reasons to explain beliefs and actions is no less scientific, no less causal and no less empirical than the postulation of any other mechanism for explaining beliefs.

Now it is true, of course, that an agent's stated reasons are not always the 'real' reasons or causes for his beliefs. Agents sometimes consciously disguise the causes of their action; other times they suffer from 'false consciousness' about why they do or believe what they do. Introspective reports of mental processes are often unreliable. Grant all that. It remains true that unless one puts forward the absurd thesis that our reasoning processes never play a causal role in leading us to believe what we believe, then one must acknowledge the fact that rational beliefs (i.e., beliefs arrived at by a process of reflection and of inference) have a different causal grounding than beliefs we come by irrationally. For instance, my belief that flying is highly dangerous and my belief that the earth is spherical were, so far as I can see, produced by radically different causal mechanisms; the evidence for that difference can be found in the fact that I sheepishly admit that I have no reasons in the one case and very convincing arguments in the other.

By asserting that rational and irrational beliefs are to be explained differently, the symmetry thesis is apparently committed to the view that reasoning processes have no causal efficacy in the production of any beliefs. Such a claim is far too strong to be taken seriously.

What has led Bloor to such an improbable conclusion? I suspect it may be the result of a narrow focusing on the features of certain rather idiosyncratic philosophical models of rationality which Bloor takes to be entirely typical of the genre. Specifically, I think Bloor has allowed his close reading of Imre Lakatos to delimit the scope of the ratonal. Recall for a moment the features of Lakatos' model of 'rational reconstruction'. In doing such a reconstruction, one uses a normative theory of scientific reasoning to decide what a scientist *should* have done, said, thought, etc. To see whether the scientist was rational, *a la* Lakatos, one examines how closely his actual behavior replicates the predictions of the model. Any divergence between the two is attributed to the 'irrationality' of the scientist.

Bloor is quite right that Lakatosian rationality is *causally* irrelevant to explaining an agent's belief. The fact that we can give a 'rational reconstruction' of, say, Newtonian optics does *not* mean that we have identified the causes of Newton's beliefs about light, for Lakatos makes no pretensions to be uncovering the real reasoning processes of historical agents. If we judge an agent's rationality by examining the reasons *we* would give for his belief rather than by examining *his own reasons,* then we have disavowed any effort to speak in the causal idiom. (Lakatosian reasons are manifestly not causes!) But what must be stressed is that Lakatos' model of rationality is neither the only, nor even a typical, example of its genre. Most philosophical theories of ratonal belief, including ones as different as Collingwood's and Hempel's, have been efforts to identify the *actual reasoning processes* whereby specific agents came to hold specific beliefs. Unlike Lakatos' model, most theories of rationality are *explanatory* in their ambitions. They maintain that rational beliefs are arrived at differently than irrational ones are, and that each calls for a different sort of causal story.

There is another relevant confusion here, specifically the linkage between the 'social' and the 'rational'. Like many sociologists (and some philosophers), Bloor does not want the sociology of knowledge precluded from handling cases of rational belief. He seems to feel that only by denying the causal relevance of rationality can he bring the rational within the scope of sociology. But there are other, *prima facie* more promising, means for keeping the scope of sociology unrestricted. One could envision a 'sociology of the rational' which would be concerned to explain why in certain cultures certain things counted as good reasons. We know, contra-Lakatos, that scientific rationality is not static, but constantly evolving. What social factors play a role in shaping the manner in which rationality itself evolves?

What I am suggesting is that even the most ambitious sociologist of knowledge need not adopt the thesis of rational symmetry in order to leave open the possibility that rational beliefs may have social origins. But the best route to asserting the territorial demands of sociology is not, I believe, by arguing that rational and irrational beliefs arise through similar or identical causal mechanisms; it is to be found, rather, by granting that rational and irrational beliefs may well have different origins and going from there to develop sociological models about rational belief formation. The subsumption of the 'rational' under the 'social', which I take to be Bloor's ultimate goal, neither calls for, nor is aided by, the monistic reductionism inherent in the symmetry thesis.

I want to elaborate on this point since I think it is at the core of the reservations many philosophers have about the strong programme. Most philosophers are persuaded that reasoned and unreasoned beliefs represent two very different doxastic states and that these differences are due to different causal ancestries for the two types of belief. Bloor could, it seems to me, readily grant that these *prima facie* differences are genuine and still seek to show that social factors were causally efficacious in both rational and irrational belief formation. Those social factors might vary in the two cases, or they might play different contributional roles, but one need not accept the symmetry thesis in order to maintain (if one is so inclined) that all knowledge is social.[21] I have a hunch that Bloor has persuaded himself that if he once grants that the causal mechanisms productive of rational belief differ from those that produce irrational belief, he will be forced to restrict sociology to the study of the irrational. But nothing of the sort is required. Rational and irrational beliefs could emerge via radically different mechanisms and the explanatory scope of sociology would not need to be curtailed at all. It is only when one imagines with Lakatos that the rational is intrinsically non-social that the sociologist need be alarmed at the suggestion that rational and irrational beliefs are the result of different causal mechanisms.

Let us pursue this approach one step further. Suppose there were a group of rational agents. Suppose we were to identify the rules by which this group of individuals 'fixes' its beliefs (to use Peirce's marvellous phrase). Suppose, further, that these rules required agents to subject prospective beliefs to certain forms of scrutiny and analysis prior to their adoption. Imagine, finally, a very different community of, say, epistemic anarchists. Their view, in so far as there is assent about these matters, is that one adopts beliefs independently of any shared cognitive policy. One may or many not have reasons for one's beliefs; one may or may not have evidence for them, etc. Now, the sociologist who wants to explain beliefs in these two societies will in both cases refer in his explanation to the belif-governing policies present in each society. That is the common core. But the 'causes' of belief in the two cases are apt to be radically different. Those in the rational society will adopt beliefs only after they have been carefully scrutinized and only after the agents have engaged in a specifically self-conscious causal interaction with relevant bits of the world. Those interactions will consitute in large measure the 'causes' of the rational agent's beliefs. By contrast, agents in our hypothetical 'Feyerabendian' community, will adopt beliefs because they take their fancy, or because they seem absurd or, more likely than not, without having any idea whatever concerning why they believe what they believe. Because these two societies adopt very different cognitive policies and thus institutionalize very different mechanisms for the formation of belief, anyone studying the beliefs of these two cultures is going to have to talk about very different causal mechanisms operating in the two cases.

Now this contrast between a 'rational' and an 'irrational' society is too starkly drawn, of course. But what this hypothetical example illustrates is that both rational and irrational behavior may have significant social components, even when the causal mechanisms productive of rational and irrational belief are very different. A programme for the sociologizing of all forms of knowledge *need* not be committed to the thesis of causal symmetry. More importantly, it *ought* not be committed to that thesis. Whether there is causal homogeneity in the production of belief is an empirical question, not to be settled prior to inquiry. Still worse, preliminary evidence, subject of course to further revision, suggests that different kinds of causal mechanisms are involved in rational and irrational actions. That fact, if it is a fact, is no obstacle to a global view of the prospective scope of sociology. But it is, and I suspect will remain, a source of grave reservations about the Bloor-Barnes version of the strong programme for the sociology of knowledge.

I do not think Bloor has yet faced up to the tension between his commitment to letting experience be our guide and the monistic and *a priori* flavor of the symmetry thesis. The same risk of contradiction appears in the work of Bloor's colleague, Barry Barnes, who like Bloor professes both to be open-minded and to be firmly committed to symmetry. The tension surfaces particularly dramatically in a forthcoming essay by Barnes and Bloor on "Relativism, Rationalism and the Sociology of Knowledge." In that essay, they insist that the sociologist should take "a completely open and matter-of-fact stance towards the role of physical, genetic or psychological and non-social causes that must eventually find a place in an overall account of knowledge."[22] But in the self same essay, even in the same paragraph, they insist upon following the symmetry thesis (there called 'the equivalence postulate'), that true and false, rational and irrational, beliefs are all caused in the same way. They cannot have it both ways. If the sociologist is going to be open-minded about the specific causes of belief, then that stance is immediately compromised so soon as he insists that we must seek the same or similar causes for all beliefs. If, as the passage above suggests, there are even 'non-social causes' of belief, what *a priori* grounds are there for expecting social and non-social causes always to be combined in a similar fashion?

Over and again, Bloor and Barnes seem to slide with ease from the claim that all beliefs are caused (which they sometimes misleadingly define as 'relativism') to the claim that all beliefs have the same sorts of causes. These are radically different doctrines -- as Bloor clearly recognized in his original articulation of the strong programme. Blurring them for rhetorical purposes only obfuscates the real nature of the chief disagreement between the strong programme and many of its critics. Few of us would deny that beliefs are probably caused; most of us would deny that all beliefs are caused in the same way. But surely all of us can agree that whether all beliefs have a common causal grounding is a matter to be settled in light of the evidence and the results of subsequent research. To presuppose at the outset a particular and highly problematic answer to that question is surely to beg the very question to which we should be seeking answers.

What I have suggested here is that it is *conceivable* that a sociological account of rational belief could be found which would not rest on a complete causal homogeneity with respect to rational and irrational beliefs. But it is one thing to grant the conceivability of such an approach and quite another to assess its current viability. The fact of the matter is that we already have a reasonably adequate framework for talking about why agents adopt the (rational) beliefs they do in the light of a largely *asociological* model of 'good reasons'. Intellectual historians and others who seek to explain human beliefs in terms of the reasoning processes of agents need make no apologies for not rooting their 'rational explanations' in sociological soil. Until such time as sociologists articulate a plausible model for the social grounding of reasoned behavior (and *no* such model is yet in hand), the common sense framework of "reasons" is not only a viable means of explaining most beliefs -- it is the only means presently available to us.

For the longer term, it seems entirely appropriate for sociologists to hope that they may eventually be able to develop a sociological model for rational belief; it is not appropriate for them to demand in advance that reasoned and unreasoned beliefs must be handled by the same causal mechanisms.

It may seem that, in accepting the thesis of epistemic symmetry yet resisting the thesis of rational symmetry as entirely *a priori* and premature, I am being inconsistent. After all, one might say, if we should leave it to future research to settle whether rational and irrational beliefs are causally homogenous, why should we not also leave it to future inquiry to settle whether beliefs in true and false theories are similarily produced? There is a crucial difference in the two cases, having to do with the *accessibility of the relevant evaluative parameter*. Because we know how to ascertain whether an agent is behaving rationally or irrationality, we can conceive of a programme of empirical research which would ascertain whether rational belief is produced in the same fashion as irrational belief. We can imagine using experience to settle the issue. But there is no parallel in the epistemic case. Because we do not know how to tell which theories are true, we cannot neatly partition theoretical beliefs into the true and the false and then proceed to examine whether they have similar or dissimilar causes. The insistence that true and false theoretical beliefs have either the same or different kinds of causes is radically untestable and thereby has no legitimate place among our basic assumptions. By contrast, we *can* conceive of ways of ascertaining whether rational and irrational beliefs are similarly caused; it is thus premature to attempt to pre-empt experience by making rational symmetry a feature of meta-sociology.

4.3 Pragmatic Rationality

It is apparently Bloor's view that the pragmatic success or failure of a belief is causally irrelevant to its persistence. He nowhere says this directly, but it seems to follow from the general formula of the symmetry thesis to the effect that evaluations of a belief are irrelevant to its explanation. Hence whether a scientific theory works well at predicting and explaining the world is supposed to have no bearing on our explanation of its fortunes. I am loath to attribute such a bizarre view to Bloor, but I do not see how the symmetry thesis can circumvent it.

It is true of course that there are social and intellectual mechanisms which manage to keep alive some very unsuccessful theories; it is equally true that some successful theories are very slow to win adherents. But in general one does want to say that most of the theories that enjoy long careers as objects of scientific belief tend to be successful at 'saving the phenomena'. More than that, we often have good reasons to expect that the success of such theories is largely responsible for their longevity. It is entirely natural, for instance, to say that Newtonian mechanics endured for so long in large measure *because* it successfully performed the tasks natural philosophers demanded of it.

I think Bloor very much needs to clarify where he stands on this issue since, of all the evaluative categories, pragmatic success seems the most natural 'explainer' of the vicissitudes of belief. If Bloor agrees, he needs to show us how the invocation of such success is compatible with the thesis of symmetry. If he does not, he should explain why the fact that a theory succeeds at doing what the scientific community expects it to has no bearing on the cognitive attitude of scientists toward it.

5. The Avowed Primacy of the 'Sociological Turn'

Ubiquitous in the writings of the so-called Edinburgh School, as well as in many other social historians and sociologists of science, is a form of argument whose structure should be carefully attended to. I call it the *fallacy of partial description*. Typically, the argument runs like this: "Science is a social activity, therefore it is best understood and explained in sociological terms." There are numerous minor variants of this (often in terms of science being a 'social construction of reality'), but its general structure can be found in dozens of writers on the subject. Bloor has several formulations of it. He says, for instance, "science is a social phenomenon so (to understand it) we should turn to the sociologist of knowledge."[23] He insists that *because* scientists undergo "education and training", "there is a social component in all knowledge."[24] The fact that science is a social phenomenon, the fact that scientists are trained by a society, manifestly does *not* warrant the claim that all or most parts of science are best understood using the tools of sociology. Only if science were exclusively a social phenomenon would the social character of science support the claim that sociology is the best tool for its study. The fact of the matter is that science is a multi-faceted process. One could as well say that science is a psychological phenomenon (considering, for instance, the role of cognition and perception in it) and thus should be studied primarily by psychologists. Equally, it is an economic and political activity. Alternatively, science is a goal-directed activity and is thus legitimately in the sphere of decision theory and operations research. Insofar as science is carried out by human animals it is presumably a biological activity. The point is that science can be legitimately studied in a variety of ways. We come closer to its essence if we say that science is a human rather than a social activity and that in its turn means that all the various sciences of man are potentially relevant.

The fact that a predicate can be truly predicated of a subject (as 'social' can of 'science') manifestly does *not* entail that our knowledge of the subject is exhausted by studying it under the predicted description. It is nothing but a bad pun to say that because science is a social activity it follows that science is best understood as a sociological phenomenon.

All this must seem obvious enough, but some of our friends in social history and sociology are loath to grant it. There is a widespread tendency to assume *a priori* that the only route to a genuine knowledge of science must be sociological. Bloor, for instance, says: "If sociology could not be applied in a thorough-going way to scientific knowledge it would mean that science could not scientifically know itself."[25] This remark sounds slightly provincial when we consider the potential 'scientific' resources open to us. It is surely entirely conceivable, although not apt to happen, that we might eventually come to have a comprehensive 'scientific' knowledge of science utilizing economics, cognitive psychology, and biology without any significant role for sociology. Again it is entirely conceivable, and much more apt to happen, that we may discover that some portions of scientific activity lend themselves to sociological analysis while others do not.[26] Whatever the ultimate outcome, it is not yet established either that all of science is amenable to sociological treatment or that any part of science is more appropriately treated by sociology than by any rival form of understanding. To argue that because science is a social activity we should view sociology as the primary tool for its investigation is like arguing that because syphilis is a social disease it is only or primarily the sociologist who can have scientific knowledge of syphilis. Insofar as Bloor grounds his case for the sociology of science on the bromide that science is a social activity, he will have to forgive some of us for being unmoved to take up the cause.

To be fair, Bloor goes beyond such sloganeering with which many social historians of science are content to rest their case. Specifically, he offers one argument which is designed to show in a perfectly general way that *all* theoretical beliefs must have social causes. I shall call it the argument from underdetermination. Bloor formulates it thusly:

> But theories and theoretical knowledge are not things which are given in our experience...this does not mean that theory is unresponsive to experience. It is, but it is not given along with the experience it explains, nor is it uniquely supported by it. Another agency apart from the physical world is required to guide and support this component of knowledge. The theoretical component of knowledge is a social component, and it is a necessary part of truth, not a sign of mere error.[27]

The core of this argument is the claim that specific theories are not "uniquely supported" by the evidence. I think that Bloor means that the observable facts of the matter never uniquely pick out one theory to the exclusion of all other possible theories. Something more than the relevant evidence is seemingly required to explain why scientists make the specific choices they do. As the passage above makes clear, Bloor believes that the 'something else' is invariably a matter of the socio-cultural circumstances and the socially defined conventions which operate on and within a specific scientific community.

The question we must ask is this: does the underdetermination of theories provide an *a priori* warrant for the claim that all theory choices have a "social component", i.e., are the result of social circumstances and conventions? At a *trivial* level, the answer is surely "yes". Scientists are educated and socialized into a certain community and they address their publications to their peers. But none of those facts entails Bloor's view that the content of theories is socially determined. To establish the latter, one would have to show that the instrumentalities scientists use for circumventing the problem of underdetermination are always social in character. There are good reasons for suspecting that they are not. One common scientific strategy for choosing between two underdetermined theories is to select the 'simpler' of the two, where simplicity is understood in terms of the number of different kinds of entities it postulates. A crude justification for this principle is that prior experience indicates that simpler theories are easier to test and evaluate than highly complex ones. On other occasions, sciientists will use other criteria (such as preservation of causality or retention of certain kinds of invariances) to circumvent underdetermination.

This is not the place to evaluate the soundness of any of these methodological principles. It is sufficient to say that until one demonstrates in a non-trivial way that such criteria of theory selection are socially caused, then one cannot argue that underdetermination *ipso facto* shows theory choice to be social in character.

But there is another, and potentially more serious, objection to Bloor's handling of underdetermination. In the abstract, theory choice is indeed underdetermined; if we imagine a scientist with access to every logically possible theory about the world, then a sub-set of those theories will indeed be empirically indistinguishable. But in explaining actual theoretical preferences by scientists are we really seeking to explain why scientist x chose theory y from the infinite range of rival theories he might conceivably have invoked? It seems to me that this is not the central explanatory problem that confronts us. Theory choice is comparative; there will typically be a very small range of genuine well-articulated alternatives open to the scientist at any given time. If one does optics in the 1830s, the choice is between the wave and particle theory of light. If one is a geologist in the same period, the choice is between uniformitarian and catastrophist theories. If one is doing electrical theory in the 1790s, the choice is between one and two fluid theories of electricity. The historian or the sociologist concerned to explain preferences in such circumstances will not generally ask, for instance, "why did Fresnel choose the wave theory of light rather than indefinitely many other empirically indistinguishable (and therefore empirically underdetermined) theories? Rather, the historian will typically ask: "Why did Fresnel believe the undular theory rather than the emission theory?" What do we gain by putting the question in this form? What we get is, to all practical purposes, a solution to the problem of underdetermination. The wave and particle theories of light are not empirically equivalent, any more than uniformitarian and catastrophist geology and one and two-fluid theories of electricity were. With respect to such theories, there were tests available to the scientific community which indicated that one member of each of these theory pairs was empirically better supported than its *extant* rivals. Underdetermination is a serious problem only when the *available* theories are *equally* well supported by the evidence. Such situations do occur in the history of science, and not infrequently. But for Bloor's omnibus argument to work (an argument which concludes that social factors are decisive in every significant doxastic act of the scientific community), he must be able to show that every theoretical preference is undetermined. He has not yet shown that to be the case.

It is entirely conceivable, of course, that social factors get into the scientific act quite independently of the elasticity allowed by underdetermination. But if so, the presence of those factors has to be independently established. All I have been concerned to show here is (a) that when theory choice is underdetermined, it does not necessarily follow that social factors are causally responsible for further narrowing the choice, and (b) that in many cases where scientists choose between theories, the circumstances of the choice are mischaracterized if regarded as cases of radical empirical underdetermination.[28]

6. Conclusion

There are at least two distinct senses in which a programmatic may be said to be 'strong.' In one sense, we say a set of theses is 'strong' if it has stood up successfully to a demanding battery of tests. This sense of strength derives from impressive evidential support. In a very different (Popperian) sense, we say that a set of assumptions is 'strong' when it is very bold, audacious and sweeping in the claims it puts forward. *Neither* attribute in isolation is worth a great deal. Neither well-tested theories of very narrow scope nor ill-tested theories of very broad scope serve our cognitive interests optimally. What we seek is theories strong in a dual sense. The strong programme in the sociology of knowledge, particularly with its incorporated symmetry thesis, is 'strong' in the second sense only. It is bold, ambitious and global in the claims it makes. But, as I have tried to show, the symmetry thesis is not strong in the crucial sense of being well supported by the evidence. Quite the reverse. Under the circumstances, one might be inclined to drop the symmetry thesis and keep the rest of the strong programme intact. Such a policy runs the risk of robbing the strong programme of its strength in the second sense, since *sans* symmetry the strong programme reduces to the relatively mild claim that beliefs are causes. What is called for, if the strong programme is to earn that designation, is some significant modification of the symmetry thesis which makes claims which are both bold and which are supported convincingly by the evidence. Until such a replacement is produced, the strong programme possesses strength in just the sense that hermeticism, mesmerism, and monism were 'strong': all sought, under the banner of 'science', to put forward theses for which there is only the most tenuous of evidence.

Department of Philosophy
Virginia Polytechnical Institute

Notes

* I am grateful to J. Ben-David, R. Laudan, A. Lugg, P. Machamer and T. Nickles for helpful comments on an earlier draft of this essay. I also owe a large debt to David Bloor, whose personal communciations have helped me clarify my ideas on these issues.

1. David Bloor, *Knowledge and Social Imagery* (London, 1976), p. 45. (Hereafter abbreviated to '*KSI*'.)
2. See, for instance: Barry Barnes, *Interests and the Growth of Knowledge* (London, 1976); Mary Hesse, *Revolutions and Reconstructions in the Philosophy of Science* (London, 1980), chap. 2; essays by Bloor, Rudwick and Caneva in Douglas and Ostrander, eds., *Exercises in Cultural Analysis: Grid-Group Analysis* (New York, 1979); S. Shapin, "Phrenological Knowledge as the Social Structure of Early 19th Century Edinburgh," *Annals of Science*, 32 (1975),

219 ff.

3. See E. Manier, "Levels of Reflexivity" in P. Asquith and R. Giere (eds) *PSA 1980*, East Lansing 1980, pp. 197-207.
4. Bloor protests the fact that philosophers "have been allowed to take upon themselves the task of defining knowledge." (*KSI*, p. 1)
5. Thus Bloor: "It is largely a theoretical vision of the world that, at any given time, scientists may be said to know." (KSI, p. 12)
6. Bloor, *KSI*, p. 6.
7. Bloor, *KSI*, p. 4-5.
8. Bloor, *KSI*, p. 4.
9. Bloor, *KSI*, p. 1.
10. Bloor, *KSI*, p. 10.
11. Bloor, *KSI*, p. 40.
12. Bloor, *KSI*, p. 17.
13. Bloor, *KSI*, p. 71. With unintended irony, Bloor urges that this uncritical acceptance of science and its methods is the only way to avoid dogmatism and ideology! He writes: "Unless we adopt a scientific approach to the nature of knowledge then our grasp of that nature will be no more than a projection of our ideological concerns." (KSI, p. 70)
14. Bloor, *KSI*, p. 1.
15. Bloor, *KSI*, p. 141.
16. Bloor, *KSI*, p. 144.
17. Bloor, *KSI*, p. 142.
18. Barry Barnes, *Intersts and the Growth of Knowledge* (London, 1977), pg. 25.
19. A general caveat is in order here. When we talk about the causes of an agent's belief, we might be referring either to the factors which first occasioned the agent to consider a certain idea (the 'context of discovery') or to the factors which caused the agent to admit the idea to his body of beliefs. Both Bloor and I are concerned chiefly with the explanation of the causes of belief in the latter sense. Those social causes relevant to the *genesis* of an idea but not to its fixation in a body of beliefs will not be discussed here.
20. For a detailed argument to this effect, see my "A

Confutation of Convergent Realism", *Philosophy of Science,* 1981.

21. In her recent *Revolutions and the Reconstructions in the Philosophy of Science,* Mary Hesse attempted to defend a "severely modified version" (*ibid.*, p. 57) of the strong programme. It replaces Bloor's symmetry thesis with the much weaker claim that rational and irrational beliefs are both "explananda of the sociology of knowledge." (p. 31). Dropped altogether is the demand that both types of belief are to be explained by the same 'sorts' or 'styles' of causal mechanisms. Given this dramatic weakening of Bloor's argument, I can only concur with Hesse's own conjecture that, "It may be felt the 'strong' thesis has now become so weak as to be indistinguishable from something any rationalist or realist could accept in regard to the development of science." (*ibid.*)
22. B. Barnes and D. Bloor, "Relativism, Rationalism and the Sociology of Knowledge", forthcoming, typescript, p. 13.
23. Bloor, *KSI,* p. ix.
24. Bloor, *KSI,* p. 28.
25. Bloor, *KSI,* p. 40.
26. Roger Trigg, a critic of Bloor's work, puts the point well when he argues: "If beliefs can be produced by a process of causation lying in the province of physiology or psychology rather than of sociology, then we must turn to those disciplines, rather than to sociology...the more it is admitted (as Bloor does admit) that the sociology of knowledge cannot tell us the whole story, the less important the subject appears and the less 'strong' its programme can be." ("The Sociology of Knowledge", *Philosophy of the Social Sciences,* (1978), pp. 289-298.
27. Bloor, *KSI,* p. 12-13.
28. I examine this issue at greater length in my forthcoming "Overestimating Underdetermination".

David Bloor

THE STRENGTHS OF THE STRONG PROGRAMME

Professor Laudan's strictures on the strong programme in the sociology of knowledge do not, I think, call for a defence of that programme so much as a thorough counter-attack. Nevertheless I am grateful to him for raising a range of objections and queries of a kind which are typical of philosophers. I shall therefore combine defence and attack. To ease the burden on the reader I shall discuss Laudan's points in the order in which he raises them. The only departure from this procedure concerns some general points from the end of his paper that I shall take up immediately. These deal with what Laudan sees as the excessive emphasis on sociological approaches and the neglect of the pragmatic success of science.

Like Laudan I think that a number of disciplines, such as biology or psychology, would have to combine with sociology if we were trying to produce a comprehensive description of all the aspects of an activity like science. I am surprised that any reader of *Knowledge and Social Imagery* could imagine that I thought otherwise. I said on page four -- and Laudan quotes the passage -- that naturally there will be other types of cause apart from social ones which will cooperate in bringing about belief. A few pages later there is, accordingly, an extended discussion of the role of sense-experience to illustrate this point. The importance of *combining* sociology with other disciplines was in fact broached in the very first paragraph of the book, and is a theme that runs through it.

The fact remains that there are good reasons for a special emphasis on the social characteristics of science, for stressing its nature as a social institution. There is nothing special about science that resides in the biology of scientists, their sensory and memory capacities, or motor dexterity. There is nothing special about the size of their incomes or the structure of their

J. R. Brown (ed.), Scientific Rationality: The Sociological Turn, 75–94.

professional organizations. There is not even anything special about their using pieces of apparatus or their taking measurements and samples. It is their goals and the interpretations they put on their interactions with the world that matter. We only begin to bring science into focus when we notice, say, certain features of its training procedure, where esoteric traditions, practices and precedents are passed on to new members. In short, it is only by examining the culture of science that we come close to the heart of that activity. To study the transmission, distribution, maintenance and change of the accepted beliefs and practices in science is therefore to study one of its most vital aspects. For this reason I can attach no sense at all to Laudan's assertion that he finds it 'entirely conceivable' that a 'comprehensive' study of science could omit sociology.

In an effort to justify his assertion my critic declares it not proven that there are any parts of science that are better treated by sociology than any other forms of study. In that case I wonder how he would analyze those parts of science which have shown themselves demonstrably affected by the presence of social interests? I will give more examples later, but as a case in point take the priority dispute between Newton and Leibniz over the invention of the calculus. The reason why sociological variables are important here is because the controversy was a symptom of intriguing connections between Newtonian philosophy and the problem of the Hanoverian succession. It was not simply a matter of the psychological make up of the protagonists. The question of priority was deliberately made into an issue in an effort to stop the future George the First bringing Leibniz from Hanover to England.[1] What was at stake was Newton's status as the Court philosopher. In order to press charges against Leibniz a delicate balance had to be struck between making out the two forms of the calculus to be *similar,* so the charge of plagiarism would stick, and saying that they were *different* so that Newton's work could be said to be superior. This is what gives the issue a connection with the detailed internal history of mathematics.

It is conventional to deplore the Newton-Leibniz dispute and hence to set it outside science proper.[2] This is the kind of response that would recommend itself to anyone who is more interested in celebrating the purity of science than in studying all its connected features in a matter-of-fact way. If this tactic occurs to my critic as a way of ruling out this example of sociological variables in 'science' I reply as follows. The connection between Whig politics and the Newtonian philosophy that surfaces in this dispute has deep and extensive ramifications in the theory of matter. Any attempt to rule out my example *a priori* would seriously conflict with the general historical picture that is now emerging of this phase of our scientific development.[3] A 'comprehensive' rather than an arbitrarily truncated picture of science manifestly *does* invoke the social.

Nevertheless, it is reasonable to insist that the most important aspects of science concern the making and testing of predictions. How does this relate to the sociology of knowledge? Isn't success in prediction connected with the *pragmatic* side of science, and that, surely, has little to do with factors of the kind investigated by sociologists? The pragmatic success of a theory is, I am sure, often connected with its acceptance and espousal by the scientific community. Given the utilitarian goals which have long informed our science this is no surprise. Of course, the connections between these successes and the *truth* of a theory is very tenuous. For the same reason the beliefs about a theory which follow from its pragmatic success are not strongly constrained. Even the acknowledgement that a theory *is* pragmatically successful involves complex judgements and is frequently a matter of dispute. Success here has to be weighed against failure elsewhere. The past history and future prospects of a theory have to be judged and compared with rivals. When pragmatic success is put in context its indications are never unequivocal nor as simple as they may seem in the abstract.

In order to justify preferences and decisions arrived at in these complicated circumstances scientists often invoke 'formal' criteria. Professor Laudan seems to think that these in some way displace or rule out social factors. As instances of such formal criteria he cites the desire that theories be 'simple', or the preference for, say, causal theories. I can only marvel at his choice of examples. Simplicity as a formal criterion has been a disaster for rationalist philosophers. There are many possible ways of defining and assessing simplicity. What, then, makes a scientist choose this rather than that definition? What gives an intuition of simplicity its credibility for this or that group? How and why might 'simplicity' be adopted in preference to some other measure of theoretical virtue which may conflict with it? All that simplicity does it to pose problems and point to the need for other kinds of explanation. It is more plausible to see 'simplicity' as an after-the-fact justification for opinions that have their real basis elsewhere.[4]

In the case of causality as a desirable or undesirable characteristic of a theory, we have the salutary reminder of Paul Forman's well documented study.[5] He has demonstrated the political significance of the rejection of causality in Weimar Germany. After Germany's defeat in 1918 there was a massive reaction against science. The immensely popular *Lebensphilosophie* of Oswald Spengler fastened on causality as a symbol of all that it rejected. Forman has shown how energetically physicists and mathematicians accommodated to this powerful intellectual fashion for denigrating causality. Even if it showed nothing else, Forman's study means that we cannot simply proceed, as my critic does, by *contrasting* professional judgements with 'social' facts, as if they could not be one and the same thing.

Professor Laudan also points out that when, for whatever reason, our choice of theories has been narrowed down to two, then experimental evidence can often tip the balance between them. Indeed it can, but Laudan's own work on Pierre Duhem will have told him that in practice crucial experiments can always be challenged.[6] So what is it that sets the scene for a crucial experiment, holding us to a given outcome or permitting us to exploit the ever present possibility of challenging it? All that Laudan is doing by *starting* his discussion at the point where there are two main contenders for the truth is assuming that social processes have already done their work and then pretending that they don't exist. Before we ever reach the point on which Laudan focusses attention we need to know how the field has been narrowed down. These are questions about the cultural inheritance of those who are conducting the crucial experiment. It is no use trying to invoke previous experimental results or the constraints of reality as a sufficient ground for this narrowing process, because this just raises all the same problems over again. We need to invoke some further processes to work in conjunction with experiment and observation. It is necessary to introduce some process such as socialization into a tradition of normal science in order to explain the constraints which limit the acceptable interpretations which can be put on the facts of experience. And if this applies to the circumstances which lead up to a crucial experiment, it applies equally to the decisions that are made about its outcome.

Providing that we do not arbitrarily choose to ignore the necessary social setting of a crucial experiment there are some extremely interesting case studies at hand. Consider, for example, Farley and Geison's account of the clash between Pasteur and Pouchet.[7] The question was: could life be produced by spontaneous generation out of non-living matter? Pouchet had conducted experiments that pointed to precisely this conclusion. Pasteur was the spokesman for the scientific establishment who challenged this belief. He claimed that minute living creatures had been introduced into the experiment and contaminated the 'non-living' matter. The appearance of spontaneous generation was thus an artifact produced by faulty procedures. This argument was consistently advanced by Pasteur even when, to our eyes, it seems very strained indeed. What Farley and Geison show is how judgements of plausibility about Pouchet's results were the outcome of social and political considerations. In the conservative France of that time the doctrine of spontaneous generation had a definite social meaning. It was looked upon as theologically mischievous and politically dangerous. The capacity to create life was the prerogative of spirit, and spirit was of course the region which housed all the ultimate justifications of the agencies of social control and political authority. To reduce life to matter was thus to attack, in a mediated and metaphorical way, the foundations of a certain kind of society.

I do not want to claim that the theoretical interpretations of experimental results is always or necessarily influenced by social factors that bear on national or dynastic politics. The question of the kind or scope of the social factors at work in a system of knowledge is entirely contingent and can only be established by empirical study. The important point, however, is that where broad social factors are not involved, narrow ones take over. The sociology of knowledge is still relevant. As well as an *external* sociology of knowledge there is also an *internal* sociology of knowledge. By this I mean that the social factors concerned may be ones which derive from the narrowly conceived interests or traditions or routines of the professional community. To see what this amounts to we may turn to the literature that deals with professional vested interests in science. This describes the concern shown by scientists in expanding the area to which their competences apply. Much that goes on in science can be plausibly seen as a result of the desire to maintain or increase the importance, status and scope of the methods and techniques which are the special property of a group. This is how Ospovat analyzes the different theories of adaptation accepted by early nineteenth-century biologists and geologists.[8] The geologists saw organisms as perfectly adapted to their geophysical surroundings. Organic changes were therefore explicable by appeal to prior geological change. Biologists, on the other hand, emphasized the variability of the connection between an organism and its environment. This made room for specially biological laws of change which would safeguard the role of their expertise. A similar set of internal professional interests lies behind the recurrent and protracted debates in botanical taxonomy described by Dean.[9] He shows how judgements on very specific issues such as the classification of *Gilia* and the number of species in the genus can be related to the interests vested in the skills and practices routine in different groups. One group will draw its boundaries according to morphological relationships, another according to patterns of gene-exchange, another according to evolutionary relationships.

Numerous other examples could be given of factors which are at once 'internal' and yet clearly related to typical sociological variables concerned with autonomy, control, prestige, status, tradition and habit.[10] I hope, however, that enough has been said for one thing to emerge clearly. To claim that scientists are educated and socialized into a certain community, and address their communications to their peers, is most certainly *not*, as Laudan asserts, to invoke a *trivial* mode of social involvement. The claim only appears trivial if it is attended by a shallow or uncertain grasp of the social processes in question. Properly understood, training, socialization and communication are deep and complex and revealing phenomena. That fact will only be appreciated, however, if we resolutely engage our accounts of knowledge with the detailed understanding of these processes that is emerging from specialist researchers.

I will now move to the more specific of Professor Laudan's objections and take them in order. They fall into three groups.

(i) The first charge, it may be recalled, is that I attack a thesis that has never been held by any significant philosophers. The thesis in question is that true or rational beliefs have, in Laudan's words, 'no explanation whatever' and that 'literally nothing causes us to believe what is true'. I agree: this thesis has never been held by anybody. But then, this is not what I am attacking. My targets in the first chapter of *Knowledge and Social Imagery* are Lakatos, Ryle, Hamlyn and Karl Mannheim. Had Laudan's book been published then I would certainly have included him in the list. What I was taking issue with was every approach that makes 'logic, rationality and truth appear to be their own explanation'. Note: 'their *own* explanation', not 'requiring no explanation at all'. it is the self-explanatory or self-moving picture of knowledge, as it is presented by these writers, which appears to me so pernicious.

Professor Laudan's quotation from my book -- about the general structure of the positions that I am attacking -- unfortunately misses out the crucial words. Where I refer to my opponents' views in terms of logic, rationality and truth being *their own* explanation, Laudan leaves a row of little dots. Had he included the missing words, this might have inhibited his 'no explanation whatever' reading of my position.

The crucial point in all this is that many philosophers, including Laudan, proceed by first evaluating a belief, and if they find it unsatisfactory they feel the need for more information to explain it. Here, and here alone, will they countenance socio-psychological causes. Thus Hamlyn[11] tells us that 'the ways in which we may perceive something can be divided into two classes -- the right ways and the wrong ways. Indeed, one way of perceiving something -- the right way -- may be distinguished from all others' (p. 11). It then emerges that 'this case provides no room for scientific explanation, since none is called for' (p. 22). Suppose we are looking at two lines of equal length in a psychological experiment; 'nothing *makes* them look of equal length' says Hamlyn, 'they just *are* so' (p. 13). A good psychologist like Kurt Koffka, who is of course one of Hamlyn's targets, does not go through this evaluative exercise.[12] He simply poses the question 'why do things look as they do?' (p. 77).

These were the contrasts that I had in mind when I spoke about a 'symmetrical' and an 'asymmetrical' stance to evaluation and explanation. This was why I chose to call the symmetrical stances causal -- remember Hamlyn's words: nothing *makes* them look equal, they just *are* so. The correctness of the perception is its own explanation. But if Hamlyn's work is a blatant piece of anti-scientific philosophizing, then so is this:

> When a thinker does what it is rational to do, we need enquire no further into the causes of his action, whereas when he does what is in fact irrational -- even if he believes it to be rational -- we require some further explanation. [P. 189][13]

Structurally, Laudan's position, quoted above, is identical in its lopsidedness to Hamlyn, Ryle or Lakatos. It makes no difference if he proceeds to call his self-explanatory reasons by the *name* of 'cause' because the asymmetry will immediately show itself by a distinction between two 'species' of cause. We will have rational causes and irrational causes to cope with. I prefer a more uncompromising notion of causation.[14]

(ii) The second objection concerns the scientific status of the strong programme. (There are, I think, some real issues here, not just a squabble about words.) Again the focus is on the symmetry postulate. How can it be scientific, asks my critic, to assume *a priori* that the same kinds of causal agent are involved in rationally and irrationally held beliefs? Should not this be a conclusion for which I produce independent evidence rather than something that I treat as a postulate? In fact my procedure was to move from a brief survey of some current work in the sociology of knowledge to a statement of the postulates. They were meant to make explicit what seemed to me good features of the practice of investigators. I cited some nine pieces of work which did not divide knowledge into two classes and channel its attention solely onto error. In other words, I did produce independent evidence, and were I writing now I would cite twice or three times that number of studies in my survey.

I think I know why Laudan may have overlooked the role that actual cases played in the formulation of the programme. He has failed to see that I am an inductivist. He consistently tries to understand my position through a haze of deductivist assumptions. For example, he asks: how could I say that the strong programme is scientific without having a clear criterion of demarcation? He wonders how on earth we could atempt to emulate science if we cannot *say* what features of it are unique? Notice the very strong verbal bias. It is as if action only seems intelligible to my critic if it can be made out to follow from stated principles. I have no such bias. The student of the piano may not be able to *say* what features are unique to the playing of his teacher, but he can certainly attempt to emulate them. In the same way we acquire habits of thought through exposure to current examples of scientific practice and transfer them to other areas. Indeed some thinkers such as Kuhn and Hesse believe that this is exactly how science itself grows. Thought moves inductively from case to case.

My suggestion is simply that we transfer the instincts we have acquired in the laboratory to the study of knowledge itself. Those like Laudan who reject the symmetry postulate are trying to stop our inductive intuitions moving from case to case. There are certain cases that have to be protected. Stop here, they say, think about this in a diffferent way, acknowledge something as special. So when Laudan asks, "How can Bloor, who protests against the assimilation of science to the sacred, be consistent in urging that the only legitimate mode in which to study science is the scientific mode?" I reply: you have missed the point. How could I be consistent if I did otherwise?

(iii) I now come to the heart of Laudan's attack. Here he offers direct counter-examples to the symmetry thesis. He argues that, properly understood, the notion of rationality shows the symmetry requirement to be misguided. Before stating the alleged counter-examples I must sound a note of caution. In the course of his argument my critic shifts his definition of rationality. It begins as a descriptive concept but finishes as an evaluative one, and in the course of this evolution it also changes its reference from being a property of an individual to being a property of a group. I think that Laudan is quite unaware of either the magnitude or the consequences of this equivocation. In order to cope with this problem I suggest we distinguish between what may be called 'natural' and 'normative' rationality. Natural rationality refers to typical human reasoning propensities; normative rationality refers to patterns of inference that are esteemed or sanctioned. The one has reference to matters of psychological fact; the other to shared standards or norms.[15]

The first step in Laudan's argument is to formulate a model of rationality. This locates the essence of rationality in the calculation of means-end relationships. A rational individual has goals and beliefs and computes ways of achieving these goals out of the material provided by beliefs about the world. (Expressing his verbal bias, Laudan also stipulates that reasons must be statable, but this point plays no significant role in the argument.) Irrationally held beliefs are those that are adopted or imposed in the actor in ways that cut across these computations. they are the result of a disruption in normal thought processes. Reference is made to the 'direct action of social and psychological forces'. So threats, bribes, brain-washsing or trauma seem to be what is in mind.

Notice how sparse this model is. It says nothing of the choice of goals or the character of the prior beliefs. Nor does it specify any standards which must be met by the means-end calculations. These would be out of place in the model. In effect all that it says is that the brain is a calculating machine. 'Irrationality' is a label for when the machine breaks down or is overridden. None of this is criticism. I am merely stating what is included in the model and what is excluded. What we have is a very elementary representation of man's natural rationality.

The reason why this is held to be a counter example to the requirement of symmetry is because the *precise* causal story to be told when the calculating maching is working will be different from when its workings are disrupted. This is certainly true. But is this really a denial of what is required by the symmetry postulate? Only if that postulate is interpreted as the demand for *identical* causal stories for discriminably different behaviour would this be a counter-example. That is certainly not what I said.

I would want to *reverse* Laudan's judgement. The computer model of means-end rationality surely *satisfies* the requirement of symmetry. The problem of understanding the working and the breakdown of a machine does not seem to me to hinge on any prior evaluation. Of course the language of 'working' and 'not working' embodies an evaluation. Nevertheless, from the point of view of the general laws and physical principles involved, whether it be the law of the lever or the laws of electronics, working and not working are just two of the possible physical states of a device.

To make the example more realistic let me go back to the procedures of experimental psychologists. This is the science of natural rationality because it develops models of our typical cognitive processes. Its theories fulfil the requirement of symmetry because, despite Hamlyn and Ryle, psychologists do not offer different accounts for, say, veridical and illusory perception. They offer single models which under appropriate conditions produce the various observed outcomes. Thus Richard Gregory's model of perception describes the brain's utilization of perspective cues and its mechanism of constancy scaling. He then shows under what conditions the brain will be tricked into computing these conclusions that we call optical illustions of the geometrical kind, e.g., the Muller-Lyer arrows.[16] If the kind of model of natural rationality proposed by Laudan were meant seriously it would be developed in exactly the same way. In fact such models which focus on means-end calculations *have* been explored in great and fascinating detail but not, of course, by philsophers. The real home of these models is learning theory and its great exponents are, in their different ways, Hull and Tolman. It would be pleasant to see rationalists acknowledging psychologists as the true leaders in their field and reading their works with the appreciation they deserve.

Now let us see what happens when Laudan's schematic model of rationality is put to use in explanations that involve groups of people. The issue that must be faced here is this. Do all cultures proceed from the unhampered working of our natural rationality, or do some belief systems depend upon its systematic distortion if they are to maintain their credibility? The empirical evidence suggests that all institutionalized systems of belief are compatible with plausible models of natural rationality. Even the belief systems that are at a maximum distance from western scientific cultures (such as magical cosmologies) do not depend on any slackening of men's native wit. Anthropological authorities of the stature of Evans-Pritchard attest to this.[17]

Professor Laudan touches briefly on these topics when he considers the case of two imaginary social groups, one of which he calls a rational society, the other an irrational society. In one, beliefs are rendered acceptable in so far as they satisfy certain shared 'rules'. The other is an 'anarchic' society in which a riot of personal preferences is permitted. Laudan's main concern is to use this example to show that even if a sociology of rational belief were possible, it would be different from a sociology of irrational belief. The very fact that the different cognitive policies of the two groups is grounded in different forms of social organization -- and hence perhaps caused by different conditions -- is taken to refute the symmetry postulate. This is clearly just a repetition of the previous mistake. Symmetry is read as identity or, in Laudan's words, 'complete causal homogeneity'.

The fact that such an example has been considered, however, seems to constitute some small concession to the procedures of the sociology of knowledge. Indeed Laudan goes so far as to say that he could envision a sociology of the rational 'which would be concerned to explain why in certain cultures certain things counted as good reasons'. Unfortunately, just at the point at which these interesting themes are being broached the exposition becomes opaque. The example of the two social groups is meant to remind us that rationality is not static and that 'social factors play a role in shaping the manner in which rationality itself evolves'. Rationality itself? How can these locutions be squared with the definition of rationality given by the model of means-end calculation? Laudan has simply left his definition behind. He has shifted from talking about natural rationality and has moved on to normative rationality.

'Rationality' now refers to the standards which say what counts as a good or acceptable reason. Surely, there is nothing in the original model which really justifies calling the anarchic society less rational than the other. Their difference simply lies in the standards, or lack of them, that are imposed on the workings of the means-end calculations of their members. Calling one society 'rational' is simply endorsing the standards used by one group and repudiating those used by the other.

For my part I am as happy to see the word 'rationality' used to refer to a set of standards as to see it used to refer to man's innate cognitive competences. I just wish my critic would make up his mind on the question. For interesting reasons ordinary usage hovers between a psychological cum causal reference and a normative cum justificatory reference, but for theoretical purposes let us keep them apart. If we do treat reasons as shared standards then we can pursue the programme of tracking back differences in standards to differences in forms of social organization -- such as Laudan's degree of 'anarchy'. But then we will find that endorsing a set of standards has no role to play in its explanation. Laudan's calling one of his groups 'rational' and the other 'irrational' added nothing to his description.

Professor Laudan does not, however, pursue these themes; so none of these points emerge. Instead he makes a sudden retreat back to his history of ideas approach. He proclaims his faith in what he calls an 'asociological model of "good reasons"'. This wretched mongrel of a concept symbolizes beautifully the muddle between matters of psychological fact and matters of shared standards that pervades his work. Since my critic declares himself impatient with anything but a naturalistic approach, the 'asociological' aspect of his model can only refer to natural rationality. On the other hand, the reference to 'good' reasons points towards a concern with normative rationality. Alas, he cannot have both at once.[18]

In order to justify the sudden retreat to a history of ideas which proudly eschews any foundation in sociology, Laudan then makes a truly remarkable assertion. He declares that sociologists have yet to 'articulate a plausible model for the social grounding of reasoned behaviour'. He repeats with emphasis that '*no* such model is at hand'. I have pondered these claims, trying to work out how they could be made with such confidence. I confess myself baffled. The fact is that there certainly *are* plausible models which show how and why our reasoning is socially grounded.

To keep the issue simple let me merely state my own preferences on this question. One way of grounding our reasoning behaviour in society is to study the way in which it is harnessed to particular social interests. The 'interest' model has been shown to work convincingly and in detail in a large number of cases. It certainly does not say everything that needs to be said, but it says a lot. The writings of Barnes, Shapin and Mackenzie have been prominent in exhibiting the workings of the interest model by relating it to historical material, both through their original research and through their interpretive surveys of the research of others.[19]

The examples described above that bear upon the role of professional vested interests in science hint at the rich supply of exemplars or model-cases that are available to guide further work. Nevertheless my critic, with his verbal and deductivist bias, may not count these as real models at all. He may feel the need for an explicit body of statements that he can point to as the single, definitive formulation of the model. Can this further need be met? Yes it can. All that is necessary is to make reference to Mary Hesse's 'network model of classification' that is derived from the work of Duhem and Quine and is developed in her *Structure of Scientific Inference*.[20] All theories and all systems of knowledge ultimately depend on systems of classification. Properly understood this model spells out in detail exactly how and why there is a social component to every single classsificatory predicate in our language.

Essentially the model is built on the fact that predicates are learnt on the basis of a finite number of instances. These are provided by teachers or authorities who must simultaneously inform and control the behaviour of the learner. The learner's task is to acquire a sense of the similarity between the cases to which he is exposed as instances of a given concept. His sense of similarity and difference must be matched to those of other language users. This involves grasping the *conventions* which are involved in the judgements about similarity and difference. One way of conveying the character of the network model is to say that it takes Duhem's arguments so seriously that it represents every act of concept application as a crucial experiment. Just as decisions are needed to fix the outcome of every experimental test, so a decision of sorts is needed about every act of concept application. And just as no hypothesis can be tested in isolation, so no concept can be used in isolation from the network of which it is a part. Since language is a shared practice that can be transmitted to new members of a social group, the decisions involved in concept application must be systematic and to a degree predictable. This is what is meant by calling them conventions. The conventional character of language is what makes the profound involvement of society a pervasive and inescapable feature of knowledge.

The particular form taken by the conventions of a classificatory network are the result of what Hesse calls 'coherence conditions'. This is the point at which it is most easy to relate the network model to case studies of the role of social interests in science. The crucial formula is that social interests are coherence conditions imposed on the classificatory network. They are factors which determine how new and problematic cases will be assimilated into the network. They reveal waht is at stake when the boundaries and scope of a classificatory term are being negotiated -- whether it be a term like *Gilia* or a term like *polyhedra*.[21] These are the factors which explain why different groups might diverge in the way that they extend and articulate their systems or networks of knowledge.[22]

In a paper called 'Durkheim and Mauss Revisited' I have set out the network model in a way that is designed to be useful to sociologists of knowledge.[23] I have done this by building it up in a step by step fashion and relating it, on the one hand, to illustrative material from the history of science and, on the other hand, to the Durkheimean Tradition in anthropology. When Durkheim and Mauss said that the classification of things reproduces the classification of men, they were nearer to the truth than their critics have allowed.[24] For present purposes the details of the argument are not as important as the point of principle. the *kind* of model I have described, and the *kind* of connection between society and reasoning that it indicates are by no means novel. All that is new is the increased quantity and quality of empirical evidence in its favour. To declare that no model is at hand which shows how reasoned behaviour is socially grounded is simply to display a lack of awareness of a tradition of work that has long been available. The failure is therefore not one of sociologists to produce such models, but one of their philosophical critics to appreciate the significance and potential of an easily accessible body of literature. But then, there is one important psycho-social law that everybody knows: that there are none so blind as those who do not want to see.[25]

Science Studies Unit
University of Edinburgh

References

1. Arnold Thackray, '"The Business of Experimental Philosophy". The Early Newtonian Group and the Royal Society', *Actes du XXIe Congres International d'Histoire des Sciences,* 3, 1970-71, 153-59.
2. For a valuable analysis of the various responses to this dispute see S. Shapin, 'Licking Leibniz' (essay review of A.R. Hall, Philosophers at War), *History of Science,* Dec. 1981
3. See for example M.C. Jacob, *The Newtonians and the English Revolution,* 1689-1720, Ithaca 1976. S. Shapin, 'Of Gods and Kings: Natural Philosophy and Politics in the Leibniz-Clarke Disputes', *Isis,* 72, 1981. S. Shaffer, Newtonian Cosmology and the Steady State, unpublished PhD thesis, Cambridge University, 1980, Chap. vii.
4. Mary Hesse, *Revolutions and Reconstructions in the Philosophy of Science,* Brighton 1980, p. 190. Mary Hesse, *The Structure of Scientific Inference,* London 1974.
5. P. Forman, 'Weimar Culture, Causality and Quantum Theory, 1918-1927. Adaptation by German Physicists and Mathematics to a Hostile Intellectual Environment', in R. McCormmach, ed., *Historical Studies in the Physical Sciences,* vol. 3, 1971, 1-115.

6. P. Duhem, *The Aim and Structure of Physical Theory,* trans. P.P. Wiener, Princeton 1954.
7. J. Farley and G.L. Geison, 'Science, Politics and Spontaneous Generation in Nineteenth Century France: The Pasteur-Pouchet Debate', *Bulletin of the History of Medicine,* 48, 1974, 161-98.
8. Dov Ospovat, 'Perfect Adaptation and Teleological Explanation: Approaches to the Problem of the History of Life in the Mid-nineteenth Century', *Studies in the History of Biology,* 2, 1978, 33-56.
9. J. Dean, 'Controversy Over Classification: A Case Study from the History of Botany', in B. Barnes and S. Shapin (eds.). *Natural Order, Historical Studies of Scientific Culture,* Beverly Hills and London 1979, Chap. 9.
10. For example see D. MacKenzie, 'Statistical Theory and Social Interests: A Case Study', *Social Studies of Science,* 8, 1978, 35-83. S. Shapin, 'The Politics of Observation: Cerebral Anatomy and Social Interes in the Edinburgh Phrenology Disputes', in R. Wallis (ed.). *On the Margins of Science: The Social Construction of Rejected Knowledge,* Sociological Review Monographs, 27, 1979, pp. 139-78. S. Shapin, 'Phrenological Knowledge and the Social Structure of Early Nineteenth-century Edinburgh', *Annals of Science,* 32, 1975, 219-43. E. Frankel, 'Corpuscular Optics and the Wave Theory of Light: The Science and Politics of a Revolution in Physics', *Social Studies of Science,* 6, 1976, 141-84. J. Harwood, 'The Race-Intelligence Controversy: A Sociological Approach. I - Professional Factors; II - External Factors', *Social Studies of Science,* 6, 1976, 369-94; 7, 1977, 1-30. R.S. Turner, 'The Growth of Professorial Research in Prussia, 1818-1848 -- Causes and Contexts', *Historical Studies in the Physical Sciences,* 3, 1971, 137-82. R.S. Turner, 'University Reformers and Professorial Scholarship in Germany, 1760-1806', in L. Stone (ed.), *The University in Society,* Oxford 1975, vol. ii, 195-531. G. Allen, 'Naturalists and Experimentalists: The Genotype and the Phenotype', *Studies in the History of Biology,* 3, 1979, 179-209. T. Brown, 'The College of Physicians and the Acceptance of Iatromechanism in England, 1665-1695'. *Bulletin of the History of Medicine,*

44, 1970, 12-30. T. Brown, 'From Mechanism to Vitalism in Eighteenth-Century English Physiology', *Journal of the History of Biology*, 7, 1974, 179-216.

11. D.W. Hamlyn, *The Psychology of Perception*, London 1969.
12. K. Koffka, *Principles of Gestalt Psychology*, London 1936, p. 76.
13. L. Laudan, *Progress and Its Problems, Towards a Theory of Scientific Growth*, London 1977.
14. For a fuller account of why I speak of a contrast between causal and teleological styles of explanation, and my reasons for imputing teleological theories to Lakatos and others who treat rationality as a self-propelling phenomenon, see D. Bloor, 'Wittgenstein and Mannheim on the Sociology of Mathematics', *Studies in History and Philosophy of Science*, 4, 1973, 173-91.
15. I have taken the term 'natural rationality' from S.B. Barnes, 'Natural Rationality: A Neglected Concept in the Social Sciences', *Philosophy of the Social Sciences*, 6, 1976, 115-26.
16. R.L. Gregory, *Concepts and Mechanisms of Perception*, London 1974.
17. E.E. Evans-Pritchard, *Witchcraft, Oracles and Magic Among the Azande*, Oxford, 1937.
18. As evidence that this confusion is endemic and does not merely represent a momentary lapse, see the following review of Laudan's book: S.B. Barnes, 'Vicissitudes of Belief', *Social Studies of Science*, 9, 1979, 247-63.
19. As well as the previous references to their work, see B. Barnes, *Interests and the Growth of Knowledge*, London 1977. D. MacKenzie, *Statistics in Britain, 1865-1930. The Social Construction of Scientific Knowledge*, Edinburgh 1981. S. Shapin, 'Social Uses of Science', in G. Rousseau and R. Porter (eds.). *The Ferment of Knowledge: Studies in the Historiography of Eighteenth-century Science*, Cambridge 1980. Also, the very valuable collection of papers in B. Barnes and S. Shapin (eds.), *Natural Order: Historical Studies of Scientific Culture*, *op. cit.*, note 9, above.
20. M. Hesse, *The Structure of Scientific Inference*, *op. cit.*, note 4, Chapters 1 and 2.

21. D. Bloor, 'Polyhedra and the Abominations of Leviticus', *British Journal for the History of Science*, 11, 1978, 245-71. This is a sociological reading of I. Lakatos' brilliant book *Proofs and Refutations: The Logic of Mathematical Discovery*, Cambridge 1976.
22. Hesse's network model was designed, in part, to attack the empiricist idea of an independent observation language. It explains why all predicates are 'theory-laden' by showing that they all depend on a negotiable 'network'. But at the same time this explains why it is correct to equate the theoretical aspect of knowledge with its social aspect. This is because examination of the model shows that the network is a set of conventions.
23. D. Bloor, 'Klassifikation und Wissenssoziologie: Durkheim and Mauss neu betrachtet', in N. Stehr und V. Meja (Hrsg.), *Wissenssoziologie-Studien und Materialen, Sonderheft 22 der Kolner Zeitschrift fur Soziologie und Sozialpsychologie*, Opladen: Westdeutscher Verlag, 1980. An English version will, I hope, be published before too long. In this paper I develop the ideas first formulated in my review of Hesse's *Structure of Scientific Inference*, D. Bloor, 'Epistemology or Psychology?', *Studies in History and Philosophy of Science*, 6, 1975, 382-95.
24. The slogan comes of course from E. Durkheim, and M. Mauss, *Primitive Classification*, trans. R. Needham, London 1963, p. 11. The materials from the history of science that I have used to illustrate the network model and, in particular, its capacity to provide a new theoretical underpinning for Durkheim and Mauss's ideas, is taken from the growing literature on Boyle and Newton's corpuscular philosophy. This work is one of the best sustained studies in the sociology of knowledge that is currently available. For a small sample see, for instance: P.M. Rattansi, 'Paracelsus and the Puritan Revolution', *Ambix*, 11, 1963, 24-32; P.M. Rattansi, 'The Intellectual Origins of the Royal Society', *Notes and Records of the Royal Society of London*, 23, 1968, 129-43; J.R. Jacob, 'The Ideological Origins of Robert Boyle's Natural Philosophy', *Journal of European Studies*, 2, 1972, 1-21. J.R. Jacob, 'Robert Boyle and Subversive

Religion in the Early Restoration', Albior, 6, 1974, 175-93; J.R. Jacob, 'Boyle's Atomism and the Restoration Assault on Pagan Naturalism', *Social Studies of Science,* 8, 1978, 211-33.

25. I realize, that my critic has shown that he is aware of some of this literature by his brief discussion of Forman, Shapin and Brown in his book. But instead of putting the cases that he considers in their proper context, he considers them in isolation and produces *ad hoc* responses to them. This is of course always possible, and is simply a technique for evading the cumulative significance of a growing body of work. This method of response is well illustrated by his reply to the valuable work that has been done on the reception of the mechanical philosophy. 'It *might* just be that Walter Charleton accepted the mechanical philosophy because -- as he explains in 400 turgid pages -- that theory was rationally preferable to its alternatives'. Yes it might, especially if you make sure that you use an historically relevant criterion of rationality. But when you have found out what Charleton thought it was rational to take into account, and when you have explained why Charleton's judgements differed from those of others around him -- and his own earlier views -- you will have engaged in the very exercise that Laudan is criticizing. If, on the other hand, there are *ahistorical* criteria that explain the changes, then why not give the rival account, rather than just say that it *might* exist?

Gary Gutting

THE STRONG PROGRAM: A DIALOGUE

Sociologist: I've been surprised at the virulence with which some philosophers have attacked the strong program in the sociology of knowledge. Particularly, it's struck me as odd that people such as Larry Laudan, who emphasize the need for the philosopher to be guided by the actual historical practice of scientists, should raise *a priori* objections to the viability of a scientific program. Why shouldn't a philosopher such as Laudan just stand back and see what the strong program comes up with? If it produces successful sociology, then he'll have to take account of its methods in his analysis of science; if it doesn't he won't have to worry about it. In either case, I don't see any room for philosophical meddling in sociological matters.

Philosopher: I think there's a serious question of whose poaching on whom. After all, the strong program proposes a methodology for sociology, and methodological questions are clearly in the philosopher's domain. Not that philosophers of science object to scientists' discussing methodology. They have to, especially when the self-identity of their science is in question; and what they have to say profits from their first-hand awareness of the content and problems of their discipline. But formulating an adequate methodology in a given context can also profit from the philosopher of science's analytic skills and broader historical perspective. So I think it's ungracious to decline a disinterested offer of aid from competent outsiders.

Sociologist: I'd be happier if I were sure the offer is really disinterested. I have the impression that philosophers oppose the strong program because they feel it threatens their role as guardians of truth and rationality. They unconsciously realize that this role depends on a particular conception of rationality

J. R. Brown (ed.), Scientific Rationality: The Sociological Turn, 95–111.

and so strike out at any attempt to formulate an alternative to this conception.

Philosopher: So you see the strong program as a view about the nature of rationality?

Sociologist: I'd rather say that it ultimately leads to one -- specifically, to a social relativism that undermines the philosopher's absolutist pretensions. But it's best to begin by thinking of the strong program as simply a sociological approach to the explanation of beliefs. Specifically, it proposes that all beliefs, whether rational or irrational, true or false, are open to the same sort of causal explanation. It certainly seems to me that it's this proposal that's the central issue in Laudan's recent attacks on the strong program.

Philosopher: I'm surely willing to agree that we should begin by discussing the strong program as a project for explaining beliefs: but I think we'll also have to later look at it as a theory of rationality. There might be a way of formulating the project for explaining beliefs that can't be faulted; but I don't think this is going to support a relativistic account of rationality. For the moment, though, let's stick to the strong program's project of explaining beliefs. You've said that the idea is to explain *all* beliefs, without paying any attention to their truth or falsity, rationality or irrationality. In other words, the claim is that a belief's epistemic status is irrelevant to the effort to explain it.

Sociologist: That's right. Unlike most sociologists of science -- especially Merton and his followers -- proponents of the strong program are not willing to concede any privileged status to beliefs that a particular social group dubs "rational" or "warranted" or "true". For too long, sociology has been invoked only to explain the beliefs that don't conform to the philosophers' standards for what people should believe and that he therefore labels irrational. Lakatos, of course, is a prime example here. He is willing to invoke sociological explanations only for that sad remnant of benighted beliefs that aren't warranted by his methodological rules. The strong program is a revolt against this philosophical hegemony that asserts the right of the sociologist to provide the same sort of explanation for all our beliefs, no matter what we think of their epistemic status.

Philosopher: This of course is the strong program's famous symmetry thesis. But this thesis seems to fly in the face of the obvious fact that there are two different ways of explaining beliefs. The first, which I'll call Type I explanation, makes essential reference to the reasoning processes that give rise to a belief. For example; Why did Herschel believe in the existence of a transuranic planet? Because he believed that Uranus moved around the sun in accord with Newton's laws and that its observed movement would not be in accord with these laws unless there were some other gravitational mass affecting it. More generally, Type I explanations have the following form:

(i) S believes p'.
(ii) S reasons in manner R and believes the results of his reasoning.
(iii) Given p', reasoning in manner R results in p.
(iv) Therefore, S believes p.

But, of course there's another way of explaining beliefs: Type II explanations, which refer only to circumstances that do *not* involve reasoning processes. These have the general form:

(i) Anyone in circumstance C believes p.
(ii) S is in circumstance C.
(iii) Therefore, S believes p.

(C, of course, is a circumstance that does not involve S's employing any reasoning processes.) This is the type of explanation we employ when, for example, we say that Newton believed he was the second person of the Trinity because of the way he had been toilet trained. The strong program's symmetry thesis pretends that all beliefs can be explained by Type II explanations. But it seems obvious that many beliefs, especially those held by scientists, are best accounted for by Type I explanations.

Sociologist: There's no doubt that proponents of the strong program have been greatly taken with Type II explanations -- maybe because it's those that undermine philosophers' paradigms of rationality. But there's no reason in principle why the strong program cannot accept both Type I and Type II explanations as sociological accounts of beliefs.

Philosopher: But if you admit these two different types of explanation, you've given up the symmetry thesis. Type I explanations explain by invoking reasoning processes and Type II explanations don't.

Sociologist: Of course there's a difference between the two types of explanation, but not one that's relevant to the claim of the symmetry thesis. What symmetry requires is that beliefs be explained without regard to their epistemic status. This means that the explanans cannot make use of any *normative* claims about the rationality of the beliefs being explained. We cannot, for example, derive 'S believes p' from an argument that has as a premise 'Anyone who is rational ought to believe p'. Now, although your Type I explanations refer to reasoning processes, there is no reason to take this reference as having normative force. That is, there's no reason to think that the explanation requires the endorsement of the reasoning process in question as valid. Rather, we can take the reference to reasons in Type I explanations as entirely *descriptive*. The explanation merely notes that the person whose belief is being explained as a matter of fact employs a certain mode of reasoning. To put it in a way you philosophers seem to like: a Type I explanation *mentions* but does not *use* (and hence endorse) the mode of reasoning in question. The explanation is good just in case the person actually does reason in this way and hence does come to believe the conclusion that follows, according to the mode of reasoning, from the premises believed at the outset. The success of such an explanation is entirely independent of views anyone may happen to have about the correctness of the mode of reasoning employed. Thus, a Type I explanation can equally refer to *modus ponens* and other currently revered modes of reasoning and to bizarre or demode methods such as Feyerabendian counterinduction or medieval appeals to *convenientia*. The symmetry thesis is maintained as long as no beliefs are explained in a way that presupposes a normative evaluation of them.

Philosopher: I'm inclined to agree with much of what you're saying. Another way of putting it might be to say that both Type I and Type II explanations are *causal* explanations, the reasons referred to by Type I explanations being of interest only to the extent that they are causes of the beliefs being explained. I wonder, however, if there isn't in Type I explanation a normative appeal to rationality at another level. After all, even if we don't endorse the modes of reasoning mentioned in the explanans, we do endorse others that are employed in deriving the explanandum from the explanans. At least to this extent there is a normative appeal to rationality.

Sociologist: Of course; but this sort of appeal has no effect on the symmetry thesis. All you're pointing out is that the sociologist of science is himself committed to some norms of rationality in his efforts to explain people's beliefs. It may be that in view of these norms the sociologist would himself evaluate the beliefs he's explaining as rational or irrational. The point, however, is that no such evaluation plays a role in the sociologist's explanation of the beliefs. If he tomorrow decides that the belief he today thinks is irrational is in fact rational, this will make no difference for the validity of his Type I explanation. Further, we should note that the sort of appeal to norms you're pointing to is present in both Type I and Type II explanations. For that reason alone, it has no effect on the symmetry thesis.

Philosopher: I think I agree with you, though I have some doubts as to whether proponents of the strong program would agree with your interpretation of the symmetry thesis. I sometimes have the feeling that they mean to exclude descriptive as well as normative appeals to rationality from the scientific explantion of beliefs. If so, the symmetry thesis would be insisting that all explanation of belief be Type II explanations; and this, I think, is an indefensible position.

Sociologist: It's certainly true that in their own efforts as sociological explanation, proponents of the strong program have been concerned almost exclusively with Type II rather than Type I explanations. But there is nothing in the logic of their methodological position that excludes Type I explanations. Symmetry requires that all explanations of belief be causal rather than normative. Since reasons, in the descriptive sense we have been talking about, can be causes, there is no reason to see Type I explanations as a violation of symmetry. On the other hand, I think there is a good reason for the sociologist to put his main emphasis on Type II explanations -- apart from the fact that these, when applied to "great moments" in science, are most disconcerting to the philosopher. Type I explanations, after all, are always explanations of beliefs in terms of other beliefs. Hence they can explain only those beliefs that are logically derivative. Now it is obvious that many of our beliefs -- and, among them, many of the most interesting and important -- are not derived but rather held in their own right. It makes good sense for the sociologist to focus on these fundamental beliefs.

Philosopher: I don't think it's obvious that we have some fundamental beliefs that aren't derived from any others. Thinking of knowledge in that way corresponds to what epistemologists call a foundationalist view, according to which there is a set of "basic beliefs" (the foundation of our noetic structure) that are somehow "self-evident" and that provide the ultimate premises for the arguments that support all our other beliefs. Many philosophers are inclined rather to a "coherentist" view of knowledge, which sees all our beliefs as logically interwoven with one another in a complex web. Each belief is supported by other beliefs and none have the privileged status of being the unsupported foundation of everything else.

Sociologist: Well, I don't want to get involved in any in-house philosophical disputes -- least of all ones among epistemologists. Let me just say two things. First, regardless of what philosophers might think is the best or only way to have justified beliefs, it seems quite clear that almost all of us in fact have a number of beliefs that are basic for us at any given time. Maybe, by some norms of rationality we should have some way of deriving them from other beliefs; but we don't. So, since sociologists are concerned with beliefs as they actually are, I think we're justified in working with a distinction between *de facto* basic and nonbasic beliefs. Secondly, even if people's actual beliefs form a coherentist web, the sociologist can still ask for an explanation of the web as a whole -- of the entire framework of a person's beliefs. This, of course, will call for a Type II explanation.

Philosopher: I'm quite happy with the distinction between basic and nonbasic beliefs, as long as we don't think of the basic beliefs as self-evident, incorrigible, or in any other way emanating an intrinsic epistemological glow. Rather, basic beliefs are just ones that happen not to be derived from other beliefs and that, nonetheless, we may be entitled to hold. This is the "weak foundationalism" that philosophers like Alston and Cornman have put forward as an alternative to both coherentism and strong, classical foundationalism.

Sociologist: Well, whatever you need to salve your philosophical conscience. All I want is for you to realize that Type II explanations have a crucial and even primary role to play in sociological accounts of our beliefs.

Philosopher: You can rest assured I realize it.

Sociologist: I'm surprised to hear you say that. Most philosophers I've talked to get very uneasy at the idea of what we're calling Type II explanations of beliefs. I can see that they would, since such explanations show that there's nothing particularly "rational" about the beliefs they explain. They're just the result of causal processes, quite independent of reason, no different from the illusions of the insane. Of course, philosophers don't mind Type II explanations of insane beliefs or even of most beliefs that ordinary folks hold about religion, politics, and so forth. It suits them quite well to think of such beliefs as irrational. But once we sociologists start on Type II explanations of, say, basic scientific beliefs held by reputable scientists, then the philosophers start to bristle. They don't want to admit that Newton's basic beliefs about gravitation were, for example, caused by the political forces of his day, since that removes Newton as a paradigm of rationality. I hate to think of how they'll respond if some sociologist of knowledge ever sets to work on the history of philosophy. In any case, your view surprises me, since generally philosophers insist on sociologists keeping away from the cows their theories of rationality regard as sacred.

Philosopher: You're assuming that the existence of a Type II explanation of a belief makes the belief irrational. I don't agree. To give a Type II explantion of a belief is simply to discover circumstances that have causally produced it. A normative theory of rationality is an account of how the rationality of beliefs varies with the circumstances that produce them. Although a Type II explanation tells us the circumstances that produced a belief, it says nothing at all about whether they were circumstances appropriate for the production of rational belief. This is the concern of a theory of rationality.

Sociologist: But if you agree that a belief has been caused by circumstances that have nothing to do with reasons, how can you claim that the belief is rational?

Philosopher: You're confusing giving reasons with being rational. If a belief were irrational whenever it is held without reasons, then all basic beliefs would have to be irrational. Then the alternatives would be scepticism or the coherentism you were uneasy with earlier. But, working as we are in the framework of a weak foundationalism, the sensible thing to do is allow that a belief need not be irrational just because we have no reasons for it. Maybe we should agree that "rational" means "held for good reasons"; but then all that follows is that the rational/irrational dichotomy is not exhaustive. There can be beliefs -- and ones that we are entitled to hold -- that are neither rational nor irrational. My own inclination is rather to use "rational" in a broader sense to apply to any belief that we are epistemically entitled to hold. Then I'm prepared to say that basic beliefs can be rational, though they owe their rationality to something other than the existence of good reasons for them.

Sociologist: I still don't see how a belief can be caused and still rational.

Philosopher: What's the problem? You don't have any difficulty seeing that a reasoned belief could be rational, even though the reasons for it are also the causes of our holding it. So there's nothing in the mere fact of a belief's being caused that makes it irrational. Do you think rather that a rational basic belief would have to be uncaused? But why should a belief that I just randomly happen to hold be rational? I rather think that would be a paradigm of an irrational belief.

Sociologist: Your view of rationality seems similar to the compatibilist view of freedom held by many philosophers.

Philosopher: It is. In both cases, the essential point is that the application of a normative concept ('free' or 'rational') is independent of the existence of a causal explanation, though it is not independent of the nature of the correct causal explanation. As a simple example, take the case of belief in other minds -- by which I mean the belief that there are other people who have the same sort of inner mental life that I'm aware of myself as having. This belief is basic for just about all of us, except perhaps a philosopher who has found a good argument for it. Moreover, the belief is surely rational when it has been produced in the ordinary way by physical, perceptual, and linguistic interactions with others of our species. We can however imagine circumstances of its formation that would make such a belief irrational. For example, it might arise in the sole survivor of a nuclear catastrophe as a psychological defense mechanism against loneliness.

Sociologist: How about an example from the history of science?

Philosopher: Well, to get a realistic example, you'd need a lot of detailed information that I don't have right now. But to suggest the sort of thing I have in mind, consider Einstein's belief that the universe is entirely deterministic, that on no level does God "play dice". If it turns out that this belief was an insight that arose in a natural way from Einstein's work as a theoretical physicist -- e.g., if it was a Gestalt emerging from his perception of various physical domains -- then I'm prepared to think of it as rationally held. On the other hand, if we found out that the belief was a compensation for psychological insecurities (or, more bizarrely, arose from brainwashing by Marxists who kidnapped Einstein in his youth), then I'd say it was irrational.

Sociologist: What you say sounds all right to me, but of course I'm just a sociologist. I imagine your philosophical colleagues would want to see you work out in a systematic way an account of how the rationality of beliefs varies with their causal origin. But sticking to our present discussion of the strong program, it seems to me that you've accepted the program's claim that all beliefs are open to sociological explanation --

Philosopher: At least *scientific* explanation. There are probably some beliefs that the sociologist couldn't explain but psychologists or biologists could.

Sociologist: Of course. So all beliefs are open to scientific explanation. Further, you agree that all beliefs are to be explained in the same way, provided this just means that the explanation is given solely in terms of *descriptions* of belief formation, with no appeal to normative concepts of rationality.

Philosopher: Right. Furthermore, it seems to me that this view provides a way of resolving the dispute between philosophers like Larry Laudan and proponents of the strong program like David Bloor. All we need is first for Laudan to agree that the explanations in terms of reasons, which he rightly insists are an essential part of our accounts of scientific beliefs, refer to reasoning processes in only a descriptive sense; and, second, for Bloor to agree that sociological explanations of beliefs may properly refer to reasoning processes in a descriptive sense.

Sociologist: I think it's also necessary for both to agree that the force of the symmetry principle is just that explanations of normatively rational beliefs need not employ a normative notion of rationality.

Philosopher: Of course. But regardless of whether the approach I've sketched mediates the disagreements, it seems to me a good basis for understanding the nature of sociological explanations of beliefs.

Sociologist: The approach is attractive, but I'm still surprised that you endorse it. It does, after all, amount to endorsement of the essentials of the strong program and, as such, requires you to accept the sort of social relativism that philosophers usually can't stand.

Philosopher: I don't see why.

Sociologist: Well, as I noted at the outset of our discussion, the methodological project of the strong program leads to a relativist theory of rationality. Proponents of the strong program have been quite clear about this. In a recent paper, for example, David Bloor and Barry Barnes say things like this: "Far from being a threat to the scientific understanding of forms of knowledge, relativism is required by it.... Because [the relativist] thinks that there is no context-free or super-cultural norms of rationality he does not see rationaly and irrationaly held beliefs as making up two distinct and qualitatively different classes of thing."

Philosopher: So, it seems, they think that the strong program for the explanation of beliefs is equivalent to a relativist theory of rationality. That's not surprising, since sociologists have traditionally seen their discipline as closely associated with relativism. Just as, according to Putnam, realism is science's philosophy of science, so we might say that relativism is sociology's epistemology. Nonetheless, the association -- not to say the equivalence -- of the strong program for explaining beliefs with relativism is not obvious. Surely, *prima facie*, claiming that we can causally explain why people believe what they do (whether rationally or irrationally) seems to be quite a different thing from claiming, with the relativist, that rationality is a purely "local" concept with no absolute validity.

Sociologist: But don't sociological accounts of why people believe what they do show that different communities have very different norms of rationality?

Philosopher: Perhaps they do and perhaps such a result has some relevance to the question of relativism. I'm not prepared to take a position on the issue right now. My point is just that the explanatory ideals of the strong program have no logical connection to relativism. Suppose that the strong program were entirely successful. This would mean that every belief ever held by anyone would have been found to have a causal explanation. Even then there would still be the task of deciding whether the circumstances causally relevant to each belief were such as to make it rational. The mere fact that a belief is caused proves nothing about its rationality or irrationality, just as, for the compatibilist, the mere fact that an action is caused proves nothing about its being free or unfree. *A fortiori*, the fact that a belief is caused tells us nothing about the more specific question of whether the rationality of the belief is absolute or relative; that is, whether it is rational or irrational only in the particular social context in which it has been produced or would be rational or irrational in any social context. So the strong program's claim that all beliefs can be causally explained does not entail relativism. Conversely, relativism obviously does not entail the strong program, since, for example, there would be no inconsistency in holding both that basic beliefs are rational or irrational only relative to particular societies and that basic beliefs are uncaused. In sum, there is no logical tie between the truth or falsity of relativism and of the strong program's claim that all beliefs are caused.

Sociologist: I think the situation is more complicated than you're admitting. Traditionally, those who reject relativism do so because they think what we've been calling basic beliefs are rational (when they are) because they have some special epistemic status. Descartes, for example, would have refuted relativism by putting forward truths such as the *cogito* that he would claim to be "self-evident" to any rational agent. Now once you introduce the idea of self-evident truths, the strong program for explaining all beliefs causally goes out the door.

Philosopher: Why do you say that?

Sociologist: Because beliefs in self-evident truths require no scientific causal explanation. Rather, we can account for any self-evident belief p held by a rational agent S in the following way:

(i) p is self-evident;
(ii) What is self-evident is believed by any rational agent who has understood and thought about p;
(iii) S has understood and thought about p;
(iv) Therefore, S believes p.

Philosopher: But why is an explanation like this inconsistent with the strong program? Can't we say that a belief's self-evidence causes it to be believed by those who believe it?

Sociologist: Perhaps. But even so "self-evidence" is a normative concept and so couldn't be properly employed in the sort of scientific explanations to which the strong program is committed. After all, "self-evident" means something roughly like "worthy of belief on its own merits" and so has no place in a descriptive, scientific account.

Philosopher: I'm willing to grant that a rejection of relativism that appeals to self-evidence is not consistent with the strong program. But recall the distinction I made a while ago between strong and weak foundationalism. What you've shown is, in effect, just that at least some forms of strong foundationalism are inconsistent with the strong program. So if we use them as a basis for rejecting relativism, we must give up that program. But weak foundationalism can also provide a basis for rejecting relativism, and it will not confllict with the demands of the strong program.

Sociologist: I guess I need to hear more about just what's involved in weak foundationalism.

Philosopher: Well, the general idea of foundationalism, as opposed to coherentism, is that there is, for any given person at any given time, a set of basic beliefs from which all his other beliefs are derived by various forms of argument. To say that these beliefs are basic is just to say that they are not derived by argument from any other beliefs. Now the strong foundationalist, such as Descartes, thinks that for a belief to be properly basic (i.e., to be worthy of belief as basic) it must have some special instrinsic epistemic properties -- self-evidence, incorrigibility, etc. that make it utterly certain and unrevisable. Roughly, the strong foundationalist thinks that some beliefs are "naturally" basic because of the special certitude we have about them. In the last twenty or thirty years, many philosophers -- e.g., Quine and Sellars -- have criticized this sort of foundationalism, arguing that there are no beliefs so privileged that they could never be rightly abandoned. As a result, they hold that there is no self-evident, unrevisable foundation of our knowledge. But that doesn't mean that there is no foundation at all. Sellars, in particular, has made it clear that he thinks there is a foundation. Although it is not absolutely certain and may change with time, there is nonetheless at each moment a set of basic beliefs to which we are entitled even though they are not derived from other beliefs and which are the ultimate premises in any justification of other beliefs. This is weak foundationalism.

Sociologist: Excuse me, but to me it seems that your weak foundationalism is itself just a version of relativism. Different societies will have different basic beliefs to which they are rationally entitled. So for us the beliefs that the murder of innocents is wrong and that trees are not persons are properly basic; but in other societies the opposite beliefs are properly basic. How is this different from relativism?

Philosopher: Of course it can be understood in a relativist way. But it need not be. We could, for example, claim that there are certain basic attitudes and beliefs that must be shared by the members of any rational community.

Sociologist: You could claim it, but the claim conflicts with the well-known fact that there are societies with conceptions of rationality radically different from ours. Philosophers should check the sociological facts before they make claims about what's rational and what isn't.

Philosopher: And sociologists should reflect philosophically on what's involved in the concept of rationality. After all, the concept of rationality is our concept, indeed a key part of our self-image. Accordingly, a rational being is precisely someone that we recognize as being "one of us". So what sense does it make to speak of notions of rationality radically different from ours? If they differ in certain essential respects, they cease to be concepts of rationality.

Sociologist: This is philosophical apriorism at its worst! You're trying to legislate the sociologist's use of concepts on the basis of your analysis of them. Don't you realize that the most important scientific development are ones that require revision of common-sense concepts? There's no reason at all that the sociologist's account of rationality has to accord with conventional ideas.

Philosopher: But it seems to me that an anti-relativist weak foundationalism can be equally derived from a sociological approach. Consider, for example, Habermas' notion that the human community is human (in a social sense) precisely because it is characterized by a set of cognitive interests that constitute the way we experience and think about the world. It seems to me that we can take Habermas' cognitive interests as *sociological universals*, derived not from *a priori* philosophical reflection on the nature of man but from a sociological theory about the nature of human communities. For example, can't we say that sociological studies show that all human communities have an essential interest in the technical control of nature? If so, and if, as Habermas claims, such an interest requires certain ways of experiencing and thinking about the world, then it follows that there are forms of experience and thought common to all human communities.

Sociologist: There might be some *de facto* empirical universals, but I wouldn't want to canonize them as "essential" to the "nature" of a human community. What have essences and natures to do with empirical science?

Philosopher: Nothing at all if we take them, as you seem to do, as Platonic ideals intuited by pure thought. But why can't we rather see empirical scientific theory as forging, out of the fire of experience, empirically based concepts of the nature of our world? Of course, such concepts are subject to revision, but to the extent that they are in fact correct, why can't they be taken as expressing scientifically discovered natural necessities?

Sociologist: I hardly think I have to tell a philosopher about the difficulties of making sense of a "natural necessity", lying between logical necessity and mere contingency.

Philosopher: Of course, but, on the other hand, remember the difficulties of giving a purely Humean account of scientific laws.

Sociologist: Well, before you head for philosophical waters that are entirely too deep for me, let me get your position clear. You say that the strong program for causally explaining all beliefs is consistent with a rejection of relativism, provided this rejection is based on weak rather than strong foundationalism. To the objection that weak foundationalism is itself a form of relativism, you reply that it need not be if we recognize the existence of sociological universals that express the essential nature of human communities. Finally, to make sense of a scientific (e.g., sociological) discovery of an "essential nature", you posit a realm of natural necessity between logical necessity and contingency.

Philosopher: That's a fair statement of my view. I'm proposing a kind of naturalized Kantianism that holds both that cognitive norms are not timeless *a prioris* but rather derive from the attitudes of existing communities and that nonetheless these attitudes define the nature of human or rational communites as such. I've mentioned Sellars and Habermas, not because I'm prepared to defend their particular accounts but rather as examples of views that accept universal cognitive norms (and so reject relativism) without denying that there are naturalistic (causal) accounts of all beliefs (and so are consistent with the strong program).

Sociologist: But nothing you've said proves that these sorts of approaches are sound or even consistent. They may just be one more example of the philosophical appetite for simultaneously eating and having one's cake. If so, relativism seems to be an inevitable consequence of the strong program.

Philosopher: Agreed. But given the lively existence of projects like those of Sellars and Habermas, I think it's precipitous for proponents of the strong program to present relativism as an obvious consequence of their view.

Department of Philosophy
University of Notre Dame

Barry Barnes

PROBLEMS OF INTELLIGIBILITY AND PARADIGM INSTANCES

Gary Gutting's contribution to this collection serves well as a symbol for the whole, which commemorates the first extended dialogue between epistemologists and sociologists of knowledge for many years. It is clear from the principal papers, by Larry Laudan and David Bloor, that the encounter did not take on the characteristics of an 'ideal speech situation' wherein an 'unconstrained consensus' could be achieved. Nonetheless it marked something of a watershed. Even the most tentative and cautious interaction represents an advance upon the frozen postures of the preceding period, and raise the possibility of more productive exchanges.

To realise this possibility, however, some unusually severe problems of intelligibility must be overcome. The two fields sustain frames of reference and forms of discourse which are difficult to align. And abstract debates on matters of knowledge and belief, truth and rationality, causes and reasons, rationalism and relativism, are consequently plagued with misunderstanding.

Let me use my own experience to illustrate the magnitude of this problem. My work has stressed the analogies between science and the alternative interpretations of nature encountered among aliens, deviants and ancestors; it has argued accordingly that those concerned with the systematic study of culture should treat science as just another typical form thereof.[1] But nowhere have I suggested that scientists are irrational, or even non-rational. On the contrary, what has led me to stress the basic equivalence of science and other forms of culture has been a sense of the reasonableness of these other forms, a point I have made clear and explicit on many occasions.[2] Gary Gutting[3], however, is just one of a number of philosophers who have found my work distinctive in "allowing for a degree of epistemological irrationality among scientists". These quoted

J. R. Brown (ed.), Scientific Rationality: The Sociological Turn, 113–125.

words (which Gutting drew from the work of another author) stand closer to its converse than to the point of view I actually hold.

Gutting, as is evident from his contribution to the present volume, has a sophisticated understanding of the problem of rationality. And my own views on the problem, whatever their faults, are clearly stated. Yet, through my initial coding and Gutting's subsequent decoding, those views were effectively inverted. Given that a major breakdown like this can occur in by no means adverse circumstances, what chances have finer points and shades of meaning of being reliably conveyed?

This may sound like the prelude to a plea for tightly formulated definitions and meticulous care in the use of language, but such policies, although no doubt of value, do not suffice: their effect is simply to shift ambiguity from one term to another. What is actually needed is the *exemplification* of the various competing perspectives, their application to specific cases: this is the only effective way of pumping meaning into the very general terms with which exchanges are currently conducted.[4]

One merit of the theory of knowledge advocated by David Bloor is that its creators have taken every opportunity to relate it to specific cases, and have conveyed the import of its central concepts more by their direct application than by abstract definition. Like Bloor I regret that the relevant materials are either not widely known or are ignored in favour of abstract programmatic statements.[5] But the particular applications which Bloor cites in this volume, although well-suited to his own argument, are not the most appropriate choices for resolving problems of intelligibility and appraisal. The Newton-Leibniz controversy, and the putative relationship of Weimar culture and quantum theory, do indeed serve as striking illustrations of the possible applications of the sociology of knowledge, but they are insufficiently revealing of its fundamental structure, and hence unsuitable as a basis for its appraisal. There is a direct parallel here with the natural sciences, where the most spectacular applications of a theory are rarely the paradigm examples employed to elucidate and justify it.

Theories need to be presented in relation to well-documented paradigm examples which focus attention upon their central doctrines and procedures and allow their advocates to display them in operation as analytical or explanatory resources. Typically such examples will be hard cases for the associated theories -- cases where difficulties reflect upon fundamental rather than upon incidental developments. A number of appropriate examples are already discussed in the literature of the sociology of knowledge[6] -- and no doubt, although he does not cite them, there are available concrete instances of the rationalist approach advocated by Laudan. But since the aim in the present context is both mutual understanding and comparative appraisal there is much to be said for a move to more neutral ground: if parties could identify a small number of well documented historical studies,[7] couched in 'observation languages' acceptable to all, and generally recognised as raising central questions, then these would provide a basis for further discussion. Let me support my point by taking a specific historical study and using it to clarify my own point of view. I do not know whether this study could serve as a paradigm instance of the kind which is needed, but it does allow me to show what I have in mind, and it is not an example particularly favourable to my own cause.[8]

The Problem of the Intermediate Fauna

Eric Mills has documented a controversy within marine biology in the period 1880-1905 concerning the existence of a distinctive fauna adapted to the oceanic intermediate layers.[9] Both the ocean bottom and its well-illuminated surface layers were generally acknowledged at this time to sustain substantial animal populations, but upon whether or not a distinctive plankton existed in the oceanic intermediate layers opinions were divided. In Europe Carl Chun's claim that such fauna did indeed exist found increasing support as the positive results of his first tows of plankton nets through intermediate depths (1887) were corroborated by a number of similar investigations. Alexander Agassiz, however, operating from the United States, was highly sceptical of this European work which he regarded as in conflict with the results of his own tows of 1880. In 1891 Agassiz embarked upon a further extensive programme of towing, and having again failed to gather samples of an intermediate fauna he expressed himself convinced of the non-existence of any such phenomenon, thereby setting himself firmly against what was rapidly become orthodox doctrine.

Mills' interpretation of these events is splendidly old-fashioned. He does not regard the various political contingencies within and surrounding marine biology as of major relevance. Nor does he see the wider theoretical perspectives of the protagonists as of central importance. Chun held that overlaps between the limited ranges of vertical migration of different animal kinds provided a 'ladder' via which food could be distributed throughout the entire ocean; Agassiz held that food could only exist in the illuminated surface layers, and at the ocean bottom where the remains of dead surface organisms would accumulate. So these men did hold theories consistent with their specific views on the immediate fauna: Chun's theory required one, Agassiz' the opposite. But Mills emphasises that these theoretical commitments do not *account for* the controversy: they were not *a priori* commitments, all-embracing visions which determined how observations would be formulated.[10] On the contrary, they emerged in the light of past experience and would have been abandoned in the face of future adverse experience: Mills particularly emphasises that Agassiz modified comparable aspects of his thinking in the light of further accumulations of evidence.

Mills himself makes sense of the controversy by focusing upon the conflicting results of empirical investigation: Chun and others trawled the mid-waters and found plankton; Agassiz trawled and found nothing. Here is the nub of the issue. Nor was this divergence simply a matter of the theory-laden accounting of empirical findings. It was, Mills insists, a matter of real states-of-affairs, of empty nets here and populated nets there.

Mills supports this claim with a beautifully worked-out analysis, of a form infrequently encountered in historical studies of the sciences. He uses our current scientific knowledge to develop a sense of the physical environment and conditions in which the nineteenth-century marine biologists worked. In particular, he shows that the specific locations chosen by Agassiz for his tows lie, and presumably lay, in areas with extremely low oxygen concentrations in the mid-waters, and correspondingly sparse populations of plankton. Consequently, even though he varied his locations and carried out extended investigations, fortune deprived Agassiz of positive results. The constitution of physical nature was such that he was fated repeatedly to fish in empty waters.

The Problem of Explanation

Let me begin my discussion of this fairly commonplace episode by noting the problems it creates for rationalist accounts of science, and in particular for Laudan's 1977 model of theory choice.[11] I do not claim that the episode refutes Laudan. My criticism is more severe still: I do not believe that Laudan provides the necessary information to allow the model to be applied to the example in the first place. How can Laudan's 'problem solving' model of rationality be used to identify which inferences were rational and which irrational in this debate, which vicissitudes of belief were in need of causal explanation and which not?

Laudan states that it is rational to prefer that theory which solves the greater number of scientific problems, but is content to rely upon accepted intuitions as to what such problems are, and what constitute solutions to them. These intuitions, however, simply do not allow the 'problems' and 'solutions' *in actual situations* to be unproblematically characterised. In the present case, as in most substantial scientific controversies, what counts as a scientific problem and what as a solution to a scientific problem, far from being self-evident features of the situation, are part of what is being contested. Did Chun's work solve the scientific problems, and leave open incidental questions about the competence, or the psychology, of Alexander Agassiz? Or did Agassiz' results raise legitimate problems which only his hypotheses satisfactorily accounted for? If one imagines that there is something in the nature of things which discriminates between these formulations, then to grapple with the actual example is much the best way of disabusing oneself of the notion. Problems as we normally understand them (and Laudan says nothing to change that understanding) are matters which people put forward as problematic. Attention to the details of actual historical examples shows how scientists may sustain alternative *prima facie* reasonable conceptions of what is problematic, and how controversies may be constituted around the alternatives. 'Success in problem solving' seems an attractive criterion for separating rationality and irrationality in scientific controversy only so long as one remains at the level of purely abstract argument.

It might be objected that rationalism remains justified whatever the merits of Laudan's particular model, since Mills' own account provides in itself a convincing rationalist explanation of the controversy. This, however, is not the case, and Mills correctly refrains from claiming that it is. Consider the situation. The Europeans knew of Agassiz' work; Agassiz was acquainted with theirs. Identical evidence was available to both sides, yet they differed in their views. Hence, if both sides were rational, something more than their rationality is required if their differences are to be explained; whereas if one or both sides were irrational, an account of what their irrationality consisted in, and what gave rise to it, is evidently required. A rationalist might suggest that it is rational to favour one's own empirical findings, *ceteris paribus,* and irrational to prefer the findings of others to them. The consequence of this, however, is that the whole of our current natural science becomes a morass of irrationality. Conversely, if it is said to be rational to accept the findings of the majority and irrational to stand out against them, one arrives at a consensual, sociological account of the growth of knowledge (although not one which I should be in any way inclined to accept). Either way, the rationalist is likely to be profoundly dissatisfied.

What then *can* properly be said of the controversy? First, it is my view that everything which Mills has to say is perfectly plausible, if far from certain. On the one hand, it is certainly *conceivable* that the events which Mills describes were significantly modulated neither by obvious political contingencies nor by the concern of the protagonists to defend their investments in existing theories. On the other hand, it is difficult to see how the events can be understood without reference to the physical environment in which they took place. If the post-tow conditions of Agassiz' nets and/or Chun's nets had been notably different there is every likelihood that the controversy itself would have been different, if indeed it had existed at all: in this sense reality had something to do with the development of the controversy.[12] Secondly, like any sequence of empirical events, the controversy is not something we can hope to describe completely or to explain finally, definitively and sufficiently: certainly a sociological analysis of the controversy can do no more than make a contribution to our overall understanding of it. But, thirdly, such a contribution must nonetheless be central, even crucial, to our overall understanding, even if all of Mills' conjectures about the basis of the controversy are assumed to be correct.

Consider first the evidential base. Nobody accepted every item in the total pool of proffered evidence as a valid item; everybody based their arguments and inferences on a sub-set of items. And the consensus eventually achieved was, among other things, a consensus upon which items should be taken into account and which ignored. Now there is no 'logical' procedure in this situation for selecting acceptable items of evidence from reports of full nets here and empty nets there, as indeed the marine biologists themselves seem to have recognised. Nor does there seem to be any scope here for talking of what 'anyman' would have actually been moved to accept simply by virtue of his psychology or his physiology. What has to be said, I think, is that the marine biologists constructed a *conventional* representation of the repertoire of valid findings in a way which is intelligible only in terms of their existing accepted knowledge, both of their objects of study and of themselves, and in terms of whatever contingent objectives and interests did actually bear upon their scientific activities. This process of constructing or negotiating conventions is described in several contributions to the sociology of knowledge literature.[13]

One of the intriguing features of the negotiations of the marine biologists in this case was the way in which different items of evidence were made out as incompatible with each other, at least for the practical purposes of theoretical inference. Filled and empty nets were treated as conflicting statements about the state of the oceanic intermediate layer. That layer, and its distinctive fauna, were each conventionally treated as one *thing* with a uniform nature.[14] This practice was accepted and followed by all protagonists: it united them in a way that their specific beliefs and theories did not. Discourse about 'the intermediate fauna' was an *institution* which pervaded the relevant sector of marine biology: the accepted mode of use of the term constituted much of the conventional character of the discourse and reasoning within the field.

Note that this reasoning was not 'conventional' *rather than* 'logical', although it was 'conventional' rather than 'logically determined'. The distinction is of central importance, and demands heavy emphasis because of the persistent tendency to see appeals to convention as denials of the reasoned character of scientific change. In this case inferences about 'the' intermediate fauna were perfectly respectable inductive inferences -- inferences such as pervade the entire culture of the natural sciences, and indeed culture generally. But inductive inferences are never determined by 'the rules of inductive logic' or anything comparable; this indeed is why crude inductivism is not widely accepted as a satisfactory epistemology. The analogies relied upon, the level of generality moved to, the scope claimed, are all endlessly problematic features of inductive inference. When inductive inference within a community takes on a stable and uniform character, it is never because the 'rules of induction' have satisfactorily imposed themselves; rather it is because a particular pattern of inductive practices has been agreed upon or institutionalised. In this case the inductive practices were inferences to the nature and distribution of 'the' intermediate fauna. The conventional pattern was focused around this category: had it not been, Chun's and Agassiz' results might have counted as separate valid findings about distinct phenomena, or indeed might reasonably have been treated in any number of other ways. It is precisely because scientific change is (inductively) logical that it must be understood in sociological terms: this way of understanding scientific change is central not because sound inference in emprical science is always liable to disturbance but because it is always inductive. Wherever (collectively accepted) inductive practices are actually encountered, they raise empirical (sociological) problems: the specific form and scope of inductive inference requires empirical explanation.

The Problem of Reconstruction

I cannot resist using Mills' example as the excuse for a brief discussion of a further point. Mills recognises that the understanding of the verbal remains of the controversy is greatly facilitated by the reconstruction of the context wherein it occurred -- including the physical environment in which it occurred. This chimes well with the prejudices of the sociology of knowledge, which takes it as a matter of course that neither the meanings of words nor the overall significance of a segment of discourse can be understood without reference to context. But it might be thought that Mills' use of our *current* science to reconstruct the physical context is nonetheless at variance with the relativist thrust of the sociological approach; indeed some historians and philosophers might themselves regard it as smacking of Whiggism.

I want to defend Mills' procedure. He recognises the desirability of obtaining a sense of the physical environment. He uses our modern knowledge to obtain a sense of that environment. *But he does not commit the Whiggish offence of using our current knowledge to infer what past scientists should 'rationally' have believed of that enviornment, or how they 'ought' to have described it.* Our current science is used merely to produce what *for us* are descriptions of that past physical environment. We must decode these descriptions out of their linquistic expression, to provide ourselves with a non-verbal apprehension of the states of affairs which appertained in the context of the historical situation. Then we can imagine how these states of affairs might have been encoded in the routine linguistic practices of the historical agents themselves. The result is good interpretative history, not Whiggism.

It may then be asked how the use of our current science in this task is to be justified, when nineteenth-century science might have been used, or the cosmological scheme of a preliterate community. There is no fully satisfactory answer to that question. I believe that our current science is the best basis for making inferences about physical states of affairs; accordingly I recommend its use in preference to any alternative, when such inferences have to be made. But I am unable, just as anyone else would be unable, to offer a secure, unproblematic, non-circular justification of my own belief. And I recognise that my belief has to be made intelligible in its turn by reference to contingent features of my own situation and of the community in which I participate. More generally: all kinds of history and sociology demand attention to context; construction of context demands theorising; theorising can never be satisfactorily justified. Accordingly, history and sociology are themselves forms of convention, customary activity: this is indeed one facet of the relativism which modern sociology of knowledge feels bound to acknowledge. I hope this very brief discussion serves to indicate that the practical correlates of that relativism are less disruptive than some might have imagined, but that they cannot be ignored nonetheless.

Concluding Remarks

As I have said, I am unsure whether my selected example is peculiarly well-suited for use as a paradigm case. At the very least, however, I hope to have illustrated the peculiar merits of exposition and argument closely related to such cases, and possibly to have shown how such exposition can make good some of the intrinsic defects of purely abstract argument. Finally, let me emphasise that although my concern here has been with the debate between epistemologists and sociologists I am confident that what I have said applies equally to those differences which are sustained entirely within the confines of the one or the other discipline. In sociology, for example, there are those who regard the views I have expressed above as plodding and restrained, and who oppose them with the claim that the world has nothing whatsoever to do with what is believed about it.[15] This is a view which I have always had considerable difficulty in understanding. The agreed use of a very few cases like that above could perhaps alleviate the difficulty, and lead to fruitful exchanges entirely within my own field.

Science Studies Unit
University of Edinburgh

Notes

1. My first systematic and extended presentation of these views is B. Barnes, *Scientific Knowledge and Sociological Theory,* London, 1974.
2. Ibid. See also B. Barnes, 'Natural Rationality: a Neglected Concept in the Social Sciences', *Philosophy of the Social Sciences,* vol. 6, no. 2, June 1976, pp. 115-26.
3. G. Gutting, (ed.), *Paradigms and Revolutions: Appraisals and Applications of Thomas Kuhn's Philosophy of Science,* Notre Dame, 1980, pp. 9-12; quotation, p. 10. A parallel misreading is to be found in N. Roll-Hansen, 'The Controversy between Biometricians and Mendelians', *Social Science Information,* vol. 19, 1980, pp. 501-517; and a rebuttal of this is in B. Barnes, 'On the Causal Explanation of Scientific Judgment', *Social Science Information,* vol. 19, 1980, pp. 685-95.
4. Cf. the semantic arguments in T.S. Kuhn, 'Second Thoughts on Paradigms', in *The Essential Tension,* Chicago, 1977; B. Barnes, 'On the Conventional Character of Knowledge and Cognition', *Philosophy of the Social Sciences,* vol. 11, no. 3, September 1981, pp. 303-35; and *T.S. Kuhn and Social Science,* London and New York, 1982. The arguments offered there in relation to the concepts of natural science and natural history also have relevance to the concepts of philosophy and the social sciences.
5. As well as the sources cited by Bloor to counteract this, see S. Shapin, 'History of Science and its Sociological Reconstructions', *History of Science,* vol. 20, Sept. 1982.
6. See, for example, those in the later chapters of D. Bloor, *Knowledge and Social Imagery,* London, 1976; also H. Collins, 'The Seven Sexes: A Study in the Sociology of a Phenomenon', *Sociology,* vol. 9, 1975, pp. 205-24, and A. Pickering, 'The Hunting of the Quark', *Isis,* vol. 72, 1981, pp. 216-36.
7. Historical studies are needed rather than the thought-experiments sometimes so skillfully used by philosophers because the debate is about the basis of *actual* credited beliefs and acceptable inferences: purely

hypothetical counterexamples are not counterexamples at all in the context of this debate.

8. An actual example which has been put to very good use in this way is the institution of the Azande poison oracle as documented by the anthropologist Evans-Pritchard. Many philosophers and sociologists -- including Polanyi, Gellner, Winch, MacIntyre and Bloor -- have used the case to exemplify their thinking about human rationality. One way of forming a tentative opinion on the possible value of paradigm cases is to compare Laudan's abstract discussion of the 'strong programme' with my own discussion of the rationality of the Azande; see B. Barnes, *op. cit.*, note 1, chapter 2.
9. E.L. Mills, 'Alexander Agassiz, Carl Chun and the Problem of the Intermediate Fauna', in M. Sears and D.M. Merriman (eds.), *Oceanography: The Past. Proceedings of the Third International Congress on the History of Oceanography*, New York, 1980, pp. 360-372.
10. Mills does however briefly indicate how theory was involved in negotiations as to what counted as intermediate fauna (*ibid.*, p. 364) and does not discount the possible importance of vested interests in theories.
11. L. Laudan, *Progress and its Problems*, London, 1977.
12. This is emphatically *not* to say that the actual states-of-affairs were inexplicable as due to technical incompetence, or anomalous circumstances, or organisms which were not the organisms of the intermediate fauna. Of course, unverbalised results can be given a great range of defensible verbal interpretations. The unverbalised results may *nonetheless* be important factors in a controversy: this is something not sufficiently acknowledged in modern sociological studies.
13. A particularly useful and relevant account can be found in D. Bloor, *op. cit.*, note 6.
14. For an extended account of the importance of *relations of sameness* of this kind see the references in note 4. The necessarily conventional character of relations of sameness, and hence of the terms with which we describe experience, is a fundamental theme of modern sociology of knowledge.

15. See, for example, Collins, *op. cit.*, note 6; H.M. Collins, 'Son of Seven Sexes', *Social Studies of Science*, vol 11, 1981, pp. 33-62; B. Latour and S. Woolgar, *Laboratory Life*, Beverly Hills, 1979.

Ernan McMullin

THE RATIONAL AND THE SOCIAL IN THE HISTORY OF SCIENCE*

If science is, at least in *some* sense, a social product, as almost everyone in these more relaxed Kuhnian times seems disposed to allow, may not its claims be shaped to some degree by social interests? And if these interests are themselves contingent features of the particular society, is there not an ineliminable contingency about even the most apparently secure scientific findings? How, then, can the rationality of science be said to transcend the society that produces it? And if it does not, can one still validly draw that distinction between *epistemic* (knowledge) and *doxa* (opinion) which has underlain virtually all thinking about science from Plato's day until our own?

Philosophers have, on the whole, tended to attribute an entirely asocial, transcendental character to scientific rationality. Bacon, it is true, did warn of the "idols" that intruded the social upon the rational, to the detriment of the latter. But he thought that, with care and persistence, these barriers to a proper rationality could be overcome. Philosophers today are not so sure, but most would argue that they *should* be overcome, that is, that the effort must be made, even if it cannot fully succeed. Sociologists of science have argued more and more vehemently in the last two decades that the extent of social influence in science (as testified to by the history of science) is far greater than philosophers seem willing to allow.

In this new controversy, philosophers have quite often been arrayed against sociologists, with historians occupying an uncomfortable middle ground. The debate itself has, indeed, taken on a troublingly territorial tone, testifying to the presence, in *this* controversy at least, of some fairly manifest social interests. Sociologists of science now lay claim to the whole areas which they take philosophers and "internal" historians of science to have usurped, while philosophers

J. R. Brown (ed.), Scientific Rationality: The Sociological Turn, 127–163.

denounce as a blatant take-over attempt the intrusion of sociologists into what had been for long an exclusively philosophic domain. And since the controversy itself is at bottom a *philosophical* one, sociologists have more and more turned philosopher in defence of new and much extended boundaries.

In the midst of all this hub-bub, most natural scientists and some philosophers continue undisturbed in their adherence to the classical logicist position: there is a formal logic of science to which the actual practice of science more or less approximates, which suffices to validate scientific claims, and which carries within itself its own intrinsic warrant, as a "logic" should. A majority of philosophers of science would, I think, argue that this position can no longer be sustained, and that the theory of science depends in some way (and here is where disagreement immediately arises) on the historical practice of science. But if this is so, then the laying out of this practice in some sort of systematic way becomes important. This is where the efforts of historians, philosophers and sociologists now converge, but where agreement seems far away.

At issue, first, is the proper way to go about "explaining" historical episodes in science, and second, the manner in which one may move from the resulting account to a theory of scientific change in general. The first question is of primary concern to historians, and the second to philosophers. At the heart of both is a problem of major concern to sociologists of science also, namely how to relate "rational" and "social" factors both in constructing historical explanations of particular events, and in inferring from these explanations to a more general account of the nature of scientific rationality.

In this essay, I want to do two things. I want to construct, first, a set of distinctions to replace the old and unsatisfactory dichotomy between "internal" and "external" historiography of science which until now has too often been adopted as the framework for discussing these issues. Second, I want to contrast two extreme views of the relating of the rational to the social, drawn from recent literature, and to suggest that the truth (as always!) must lie somewhere in the middle. I mean to focus on the natural sciences; the social sciences would require special treatment, since the issues can become considerably more complicated there.

1. Epistemic versus non-epistemic factors

The scientist is making a claim about the world and is attempting to provide evidence for the claim of a sort that others will find acceptable. He has thus to be ready, if necessary, to provide some sort of justification for the second-order principle he is implicitly using which counts this evidence as *appropriate* to the claim he is making. This (and, of course, much more) is necessary to the human activity we call "science." If something does not qualify under these conditions, it may be a perfectly worthy activity, but it has to be called something other than "science."

But this condition is broader than is often realized. The criteria that constitute something as a valid knowledge-claim about the world in the eyes of the person proposing it may differ from those that a historian or a philosopher later discussing his claim would accept. The set of factors which enter into an episode in the history of science as evidence or as principle, I shall call "epistemic." They are constitutive, in the mind of the scientist (we are talking here about science, but the term could be used in other knowledge-contexts also), of its claim to count as knowledge about the world. They thus serve to "explain" the outcome in the quite precise sense that for the agent involved, they are the knowledge-claims that actually serve as warrant for the claim advanced (theory proposed, etc.). They give what the agent sees as reason to believe that the conclusion is an appropriate one. It may be quite difficult, of course, to decide in a given case what *did* count as warrant for the scientist involved. Even the scientist himself may be somewhat unsure of the exact balance of reasons that led him to adopt a particular conclusion. But the criterion for deciding whether something counts as an epistemic factor in a given case is reasonably well-defined. An epistemic factor is one which the scientist would take to be a proper part of the argument he or she is making. Other factors affecting the outcome arrived at can be described as "non-epistemic."

Some examples may help to anchor this distinction in the historical practice of science. Reports of experiments, verifications of predictions, background theories, claims of inconsistency, are the sorts of epistemic factor we are accustomed to in standard "internalist" accounts of the history of science. They are the sorts of considerations that would weigh with a contemporary scientist who might be repeating the earlier work or discussing it in a science course. But we have to look more widely than this. "Metaphysical" principles, for instance (i.e. principles whose warrant is claimed to be broader than, or different from, that of science) have always been of importance in understanding why scientists act as they do.

When Descartes equated matter with extension in his mechanics, thus rendering it incapable of recognizing variations in density, he relied upon an epistemological principle that for him took precedence: the concepts of a proper science have to be expressible in the form of clear and distinct ideas. *Extension* was the clear and distinct idea, par excellence, since it defined the domain of geometry. The notion of a "quantity of matter" that could vary within the same volume was unacceptable because it could not be formulated in the way that the principle required (although, as we know, Descartes was not always quite consistent in requiring matter to be characterizable in terms of volume only). Likewise, he ruled out all forms of attraction at a distance, despite the authority this notion had already gained through Kepler's attempts to explain planetary motions. And his reason was once again the same: *attraction* did not measure up to the standard of a properly scientific idea. If we are to understand Descartes' science, it is obviously necessary to grasp how important for him this principle was which required "clear and distinct" formulation of all permissible scientific concepts in advance of their employment in science. We may reject this principle as inadmissible rationalist epistemology, but the fact remains that it did serve as a crucial epistemic factor in Descartes' own thinking.

It is easy to find other examples. There is, for instance, the familiar story of the influence of German *Naturphilosophie* on both physics and biology in the nineteenth century. But lest it might be thought that such stories belong to the past, and the epistemic factors of this sort play no part in contemporary science, the recent history of quantum theory can serve as a useful reminder. The long-standing disagreement between Bohr and Einstein regarding the acceptability of the "new" quantum theory of 1924-27 was rooted in a deep metaphysical divergence. Einstein regarded the world as deterministic, as governed by specifiable causal factors that in principle, at least, ought to allow precise prediction of single events. This principle had proved itself not only in classical mechanics but also in the new relativity theories, and had acquired (in Einstein's view) the status of a regulative principle for mechanics generally. He was convinced that the universe displayed a fundamental coherence and order, and that the task of the scientist was to discover that order.[1]

Bohr, on the other hand, was deeply influenced by those nineteenth-century philosophers (including his own compatriot, Kierkegaard) who had challenged the adequacy of the mechanistic world-picture and the materialism associated with it. He had, thus, *antecedent* reasons to accept a theory which seemed to imply that nature is indeterministic at the atomic level. In his view, this was not just a temporary restriction; he was convinced that there were no deeper causal laws still to be discovered which would restore a determinism of the classic sort. Since the issue was still open at this point as far as quantum theory itself was concerned (one could consistently try to find the hidden parameters needed to supplement the quantum account), his conviction was reached on grounds that were partially philosophic in character, as he himself makes clear in *Atomic Theory and Description of Nature.*[2]

Metaphysics and epistemology are not the only additional source of epistemic factors. Many scientists in earlier centuries made use of theological premises in arriving at conclusions about the nature of the world. Geologists in the eighteenth century, for instance, commonly assumed, on the testimony of Scripture, the occurrence of a great flood as the main influence in shaping the earth's surface.

Biologists up to Darwin's day usually took for granted that the universe was brought about by an intelligent Creator; such a Creator would be likely to adapt the natures of living things to make each kind capable of survival in the niche in which it was placed. This was a premise in every discussion about adaptation. One might easily misunderstand the common use of adaptation and other similar features of organic nature as an "argument from design" for God's existence, and suppose that the existence of a Designer appeared only as *conclusion* in biological discussion of this type. In fact, it was already there as *premise,* affecting, for example, estimates of plausibility for alternative modes of explaining the manifest adaptations of the organic world. Further, the separation of man from the rest of organic nature which for so long made it impossible to suppose a common origin for both in a single evolutionary scheme, was rooted in a theological account of human spirit and of human destiny.

It is important to see in what sense premises (premises resting on Scripture, on the authority of the Church, on the tradition of a religious community, or the like) may be called "epistemic." Those who relied on them took them to be a proper part of the argument-structure. They might or might not have regarded them as properly "scientific," depending on how broadly they were willing to extend the domain of "science" (or natural philosophy, as it would then have been more likely to be called). But they would have resisted any suggestion that it was improper to make use of such premises in coming to a conclusion about Nature. An *"episteme"* about Nature in their view could not properly leave theological premises out of account.

How cleanly does the knife cut here? Suppose one could discern the influence of such premises on a scientist who, nonethless, never explicitly acknowledges this influence and even tends to draw back from any explicit endorsement of its legitimacy in a scientific context. (It would be easy to find examples of this in nineteenth-century debates about evolution). The crucial question would be whether there is good reason to think this premise did influence the scientist's reasoning in a way that *he himself* would think legitimate and ultimately defensible. Suppose, further, that the scientist's own understanding of the theological premise he is relying on is that it rests ultimately on an act of faith. Could it still qualify as "epistemic"? It could, provided it appears reasonable to him to rely on premises of this sort, i.e. he takes religious faith to be an appropriate avenue to truth about the world. Thus, the premise functions for him as a step he is willing to acknowledge in a larger argument. Suppose, finally, that the historian who later attempts to understand how the scientist arrived at his conclusions claims to show that the purported theological (or metaphysical) premise was not the "real reason." This is where the borderline may be crossed, and where the "border" itself may be hazy.

Why take so much trouble about seeking a "border" here in the first place? The conventional external-internal distinction sets everything that is not "internal" to science (as the *historian* understands science) as "external." Thus, philosophical and theological considerations are treated as "external." And from that it is only a small step to seeing them inevitably as symptomatic of underlying socio-psychological factors at work, in a way that "internal" considerations are not. There is, of course, a positivist presupposition at work here, an assumption that since metaphysical and theological assertions cannot be taken seriously as knowledge, they have to be explained away in "externalist" terms. Whereas in the case of science, since it is *real* knowledge, one could dispense with externalist modes of explanation if one wished.

As an example of this approach, take Paul Forman's well-known essay linking "Weimar culture" with the favorable reception of quantum theory in Germany in the 1920's,[3] Forman takes the philosophical considerations which influenced people like Bohr and Heisenberg to be external to science and *therefore* in need of social explanation. He sees them as manifestations of culture in a way that the formal theoretical features of quantum theory were not. The impression is given that they are fully explicable in social terms, that Bohr's prior preference for indeterminism, for example, is to be understood ultimately, not by means of the reasoning Bohr could bring in support of it but by the social causes operating on him which "causes" him to hold that belief. This implicit asymmetry between science and philosophy is unacceptable; we shall return to it below.

A related reason for preferring to cut between epistemic and non-epistemic is the common tendency among historians in the "externalist" tradition to treat philosophic and theological considerations as "values," as expressions of preference, to be explained therefore in socio-psychological, rather than in cognitive, terms. In his recent book, *Between Science and Values,*[4] Loren Graham takes the positivist dichotomy between fact and value all too seriously, and tends to set science and philosoophy on opposite sides of the cut. This then leads him to assume that different modes of historical explanation are appropriate to the two. "Values" (which he takes to include both philosophical and theological assertions) are expressions of the culture, whereas "facts" are testable by experience. He admits that value-considerations may enter into science proper,[5] but on the whole seems to set science on the side of "fact."

His main concern is with the influence of science upon "values"; the reverse influence (the concern of externalist historians of science, he reminds us) he seems to regard as properly explicable in terms of politics and sociology. But it is important to recognize that philosophical and theological considerations operate as *reasons.* Even though we can (and should) ask what social factors may help us understand why these considerations *do* function as evidence for the particular scientist, if our goal is to achieve an understanding of the historical episode itself, the interrelationship of premise and conclusion must be explored to the fullest extent possible. And that demands a distinction between the modes of explanation appropriate to the epistemic and to the non-epistemic.[6]

Epistemic factors operate, then, as reasons rather than as causes. They are propositional in form. It is not the experiment or the belief-state which counts as an epistemic factor in the reconstruction of an episode in the history of science, it is the proposition in which the experiment is described or the belief expressed. I do not want to get involved, at this point, in the tangled debate between those who affirm and those who deny that reasons and causes must be kept distinct. It is enough to recall that belief in something *as* a reason obviously itself functions as a cause, so that it is easy (if one wishes) to correlate epistemic factors with causes in any particular historical reconstruction.

This may help us to resolve one further ambiguity fostered by the older dichotomy between the "external" and the "internal." Consider, for example, the "Merton thesis," that Puritanism fostered the growth of science in seventeenth-century England. This has been taken to be the very paradigm of an "externalist" explanation. Yet a Puritan scientist might well remark that as a Puritan, he had a perfectly good *reason* to adopt an empiricist-instrumental approach to Nature. From the fact that it was a sociological type of historical inquiry that first suggested that Puritans as a group played a disproportionately large role in the advance of science in England around 1650, one cannot conclude that Puritanism has to be treated as a social factor of a non-epistemic sort in explaining what happened. Social factors are quite often mediated by reasons. Puritan theological principles could very well have served as epistemic factors in this case; such, at least, is the suggestion that the results of the sociological analysis prompts. Membership in a social group which shares some body of beliefs automatically indicates what *epistemic* factors are likely to come into play in the cognitive activities of members of the group. Whether these beliefs would count as reasons *outside* the group is of no concern if the question to be decided is why science developed where it did when it did. But, of course, this issue can become a major concern in *other* contexts.

2. Standard and non-standard epistemic factors

The historian *may*, of course, wish to separate out what would nowadays count as "science" from that which would not. Forman and Graham are implicitly telling us what *they* think of the use of philosophical or theological considerations in a scientific context. They are *evaluating* the arguments advanced or the research strategies followed in the historical cases they are tracing, and separating out what appear to them as the "standard" epistemic factors, the factors that in their view constitute the episode as a "scientific" one. To construct a history of *science* (they might say), one has to decide what *counts* as science, and this inevitably requires the introduction of evaluative norms.

The decision as to whether a particular epistemic factor is "standard" or not depends upon the notion of "science" employed by the person making the distinction. The boundary between standard and non-standard will vary as notions of scientific rationality vary. The distinction presupposes a set of values on the part of the person using it; it may be imposed on the grounds of current usage or it could be proposed as a reconstruction of what it was that made science progress in the way it has. It is the influence of "standard" factors that is presumed to account for the progress of science, however that progress is understood. It is the influence of non-standard factors, as we have seen, that strains the conventional external-internal distinction.

But why should a *historian* wish to focus on standard factors only? One innocuous-sounding answer is that this must be done in order to define those aspects of human history that are of interest to a historian of science. Thus, the seventeenth-century historian of science may lay aside alchemy or hermeticism on the grounds that these are not part of the mainstream that would later be seen to constitute science.[7] But what if this distinction would not have been made by some of the people whose work the historian is trying to understand? Can their work be historically understood if one prescinds entirely from non-standard domains of this sort? The answer is obvious from recent work on seventeenth-century science, and it is negative.

A second answer we have already seen and dismissed is that standard factors require a different sort of historical explanation (one in terms of the rational) from non-standard factors (one in terms of the social). Is there *any* context, then, in which this sort of distinction could be legitimate? To answer this, one further precision may be needed.

3. Rationality: implicit and imputed

The "rationality" of a particular episode in the history of science can be construed in two rather different ways. One might ask what form of rationality the scientist himself embraced. What principles of argument did he follow? What counted as evidence? What criteria did he appear to use in evaluating explanations? The list of questions is a familiar one. The aim is to discover what sort of norms actually guided the scientist in his research. This is sometimes quite difficult to discover.

One reason is that scientists usually do not lay out these

principles in explicit form. And when they do, they quite often misdescribe what they are doing. It is notorious how far some of the pioneers of science (Descartes and Newton come to mind) departed in their own practice of science from their manifestos on method. But our concern here is not with the scientists' declarations except insofar as these seem to reflect the rationality inherent in their research, the processes of argument that actually constituted it. To discover what we shall call the *implicit* rationality of a piece of scientific work, requires one to attend to every clue as to what epistemic factors *actually* influenced the scientist.

The historian who is trying to explain how that piece of science occurred will be concerned with its implicit rationality. He wants to trace the style of argument, the evidence brought, the weight given it. And since he wants to understand the decisions made, he will include all relevant epistemic factors, standard and non-standard. He will, of course, pay attention to non-epistemic factors also to the extent that they bear on the course or outcome of the scientist's inquiry. This does not fit neatly into either "externalist" or "internalist" categories, as these have usually been understood.

A different approach is to focus on those factors of the historical record which appear (to the later inquirer) to constitute a piece of inquiry as science, and to impute a rationality to it which the scientist himself might or might not have recognized. This *imputed* rationality (as we shall call it) is claimed to be consitutive of the earlier inquiry as science, but it abstracts from non-standard epistemic factors as well as non-epistemic factors that, in the judgment of the historian, did not contribute to the *scientific* status of the results obtained. This is the way in which much "internal" history of science has been written, and it is perfectly legitimate, provided it be made clear what is aims are.

It is not "history" in the broader (and more familiar) sense of an explanation of human action. It does not purport to tell us why the scientist proceeded in the way he did, what the influences were upon his action, what his own self-perception was of the inquiry he was carrying out. To single out the imputed scientific rationality of a historical segment of human reflection and action is to isolate in an abstract way a set of principles, observations, explanations, and to determine the relations between them in the light of an "acceptable" theory of science.

To illustrate the two types of rationality, we might take a familiar illustration, Newton's *Principia*. A historian who is concerned with the implicit rationality of that work will note that Newton claims that his method is inductive, proceeding by generalizations from singular observations. He takes the axiomatic section with which the *Principia* opens to rest on inductive evidence of this sort. The implication is that each of the three "Laws" is based on an inductive generalization from observed instances of the operation of "force." Newton lays no particular stress on the fact that Kepler's laws can be derived, approximately at least, from the axiomatic section, with the addition of the inverse-square law of force. Rather, he appears to treat this as an *application* of a mechanics which already has a secure inductive basis. The success of this application goes to show that the law of force governing the planets is inverse-square with distance; he does not appear to take it to be a major reason for accepting the axiomatic section itself. The hypothetical status of the notion of force introduced by the eight Definitions and three Laws of the axiomatic section is obscured by the way in which he separates the "mathematical" and the "physical," claiming that he can prescind from "physical" issues (e.g. the status of the "attractions" that are claimed to explain the motions of the planets), and focus on "mathematical" ones which leave the ontological status of his notion of force in abeyance.

A historian who is trying to establish what the warrant for the *Principia* "really" is might, on the other hand, lay aside the inductivism of Newton's protestations, and note that the three Laws are not really inductive in their basis (i.e. they are not based on generalizations from specific sets of observational data). The success of the *Principia* might be held, first, to lie in its ability to predict the planetary orbits and the periods of the terrestrial pendulum within acceptable limits of accuracy. These are not so much *applications* as *tests,* and the primary warrant for the *Principia* is that it survived these three tests. The axiomatic section might also be commended for its coherence, and for the intuitive "fit" between the Definitions and our ordinary experience. But Newton did not begin from them; he arrived at them, slowly and with many revisions, *after* the successful preliminary derivation of the planetary orbits. The *Principia* is, thus, hypothetico-deductive in form, and it rests in a crucial way on the introduction of a very complex explanatory concept, *force,* which is defined, not in inductive terms but by means of the axiomatic section, taken as a whole.[8]

A different sort of rationality is being imputed here. There is no suggestion that this is what *Newton* had in mind. It is not presented as a plausible reconstruction of what Newton believed himself to be doing. So it is not an explanation of the historical event, in the normal sense of historical explanation. What it *does* purport to explain is why Newton's mechanics worked as *science,* given some standard conception of science.

One other context in which this distinction between two types of "historical" rationality must be observed is that of *philosophical* attempts to ground a theory of scientific rationality on historical case-studies. On which of the two ought such a theory rely? If a survey is made of the implicit rationality of a series of case-studies, there will be differences between them reflecting the contingencies of the different cases. How are these differences to be composed in order to achieve a composite description? If we find one scientist relying on inductive generalization, another emphasizing aesthetic criteria in choosing among theories, another involving a prior philosophy of nature, how are we to assign weight to these different "rationalitites"? We might ask which of them was most "successful" in terms of some previously agreed-on general criterion of "success." Or we might ask whether there has been a historical development in rationality, just as there has been in science itself, as scientists learnt through their predecessors what strategies work best. (Once again, some agreed-upon notion of "working best" will have to be specified.) There are many problems here; my intention is simply to indicate this as one possible way of proceeding.

The other alteranative is to test a theory of scientific rationality against the imputed rationality of a selection of major episodes in the episodes in the history of science. Or to put this in another way, it is to see whether these episodes can be reconstructed in the light of a proposed theory of rationality. The delicate question here, of course, is how this sort of "reconstruction" is to be accomplished. One has to pick out, first, some episodes which are indisputably paradigmatic of "good science" and then decide whether one's theory of science would account for their success. One might, for instance, choose the original Copernican theory and ask whether on one's *own* reconstruction of the relation of evidence to hypothesis in this case, one could say that the Copernican view was superior to the Ptolemaic alternative (as it indisputably turned out to be in the long run). Note that this is not to ask what criteria led *Copernicus* to prefer it. Few today would accept the Pythagorean philosophy of nature which apparently gave him confidence that the sun must be central to the planetary system. So to impute an evidence-structure to the Copernican theory in the light of a later account of scientific rationality (on the assumption that this theory was, in fact, a "better" theory) is *not* to propose a historical explanation of how it all happened. But it often *sounds* as though this is what is being done. One can, thus, sympathize with the uneasiness that historians often feel with such "reconstructions" (or, indeed, with "internal" history of science generally). Yet this sort of reconstruction is not *necessarily* to be excluded; what is necessary is to make sure that no one takes it to be historical explanation in the ordinary first-level sense.

One familiar illustration of the perils of such "reconstruction" is to be found in the work of Imre Lakatos.[9] He attempted to support his "methodolgy of scientific research programs" by drawing upon detailed case-histories in the natural sciences. In order to do this, he had to "reconstruct" these episodes in small ways to make them fit the MSRP, giving rise to an obvious suspicion of circularity.[10] His defense was that he could legitimately *impute* the MSRP rationality provided that the theory-evidence relationships figuring in the episode lent themselves to reconstruction in those terms. This defense could conceivably work, if the reconstruction were sensitively carried out. The measure of this would be the plausibility of the evaluation of the Copernican "research-program," say, in terms suggested, not by Copernicus, but by Lakatos' MSRP.[11]

But where Lakatos made an irretrievable error was in

suggesting that to the extent that the reconstructed episode did not conform with the *real* historical sequence (i.e. to the extent that the *implicit* rationality of the episode differed from the *imputed* rationality derived from the MSRP), one had to suppose that socio-psychological factors were responsible. He committed himself to holding that the historical actors *would* have followed a MSRP rationality if non-epistemic factors had not interfered. But this was not necessary for the main case he was trying to make which was that the case-studies he was citing could be used to support the MSRP in a quasi-inductive manner.

To make this case, one does not have to rest on the *implicit* (actual historical) rationality of the episodes considered. But if one does, one will have to account for the inevitable divergences between these episodes and the ideal of rationality proposed. One might do this by simply recognizing that the actual could be expected to depart from the ideal in ways that involve epistemic and non-epistemic factors in historically conditioned ways. Lakatos wanted to argue a stronger quasi-rationalist thesis, namely that the MSRP ideal of rationality has been implicit in the historical development of modern science ever since its explosive period of "success" began, and that to the extent it was *not* present, this could in principle be accounted for in non-epistemic socio-psychological terms. This is a thesis about the proper *historical* explanation of all these episodes, and as such is vulnerable to the criticism that it is an implausible piece of "modern-mindedness" to suppose that a modern ideal of rationality would have been employed by the scientists of the seventeeth century had they not been prevented by socio-psychological causes from doing so.[12] Lakatos' rationalism, his confidence in the MSRP as the underlying *implicit* rationality, where not "prevented," of the history of modern natural science, thus led him to adopt a strategy that could not fail to offend historians and sociologists of science alike.

4. The presumption of standard rationality

What went wrong with Lakatos' reasoning was the assumption that in order to employ historical case-studies as a warrant for his MSRP, he had to argue for the effective presence of the

MSRP ideal of rationality throughout the history of modern science. This defines one extreme manner of relating the rational to the social in regard to the history of science: to suppose that to the extent (and only to the extent) that an episode from the history of science conforms to some standard contemporary account of scientific rationality, non-epistemic factors can be left out of account entirely. One presumes a standard of rationality in reconstructing past science, and only where this presumption clearly fails, does one look for an explanation in sociological or psychological terms. The "presumption of standard rationality" (PSR), as we shall call it, has two implications: one is that insofar as an episode in the history of science does *not* conform to the "standard" account, non-epistemic factors are to be held responsible; the other is that *only* to the extent that case-histories do not conform can explanations in socio-psychological terms properly be sought.

In his book *Progress and Its Problems*,[13] Larry Laudan is more explicit in defence of a PSR.[14] It is crucial, he thinks, as a demarcation between philosophical and psychological analysis:

> Until the rational history of any episode has been written (and that by using the best available theory of rationality), the cognitive sociologist must bide his time; to do otherwise is to abrogate the [PSR].[15]

Where exactly does the line of cleavage come? Laudan is critical of those who would suppose the influence of metaphysical or theological considerations on scientists of the past to be a sure sign of the operation of (non-epistemic) socio-psychological factors. (This is the position we attributed above to Forman and Graham; Laudan takes Kuhn to be of their number, but Kuhn would, I think, deny this.) His own standard of rationality is defined by the problem-solving model developed in his book, to which in effect he assigns a transhistorical status.

This standard would allow the possible legitimacy of the influence of, say, theological considerations on science at some epoch in the past (though not in the present, so there is a second evaluative, criterion at work), provided that the theological system in question be an "effective problem-solver."[16] So it is (in our terms, at least) a more nuanced criterion than that of those who effectively collapse the category of the epistemic to that of the "standard rational." Nonetheless, there are some obvious difficulties in the claim that if Newton's theology, which admittedly influenced his mechanics in some crucial respects, was (in whose terms?) lacking in problem-solving effectiveness, one would have to look for socio-psychological factors to explain its influence on Newton's scientific work.

Laudan formulates the consequences of his PSR in a particularly clear way:

> (1) The domain of possible belief-situations for sociological analysis is restricted to those in which agents accept beliefs or weigh problems in ways incompatible with what rational appraisals would suggest; (2) the sociologist of knowledge must be able to show that the theory of rationality to which he subscribes (in order to determine what cases might be sociological) is the best available one.[17]

The first objection to a PSR when defined in this way is aptly brought out by a consideration of (2) above. If we discern a failure of rationality in the beliefs or arguments of a scientist, we are entitled to ask whether some non-epistemic factor might not have been responsible. But a failure of rationality in whose terms? Surely not in ours? Surely not in terms of the "best available" (modern) account of scientific rationality? To suppose that there is an underlying "standard" rationality, even of the most general sort, failures in which in any culture at any time are indicative of the intrusion of socio-psychological factors, is hazardous, to say the least. The issue, recall, is not whether an ideal of rationality can ever be imputed to episodes in the history of thought in order to judge their rationality, or to provide indirect support for the ideal itself. We have seen in the previous section that under certain restrictions, this *can* be legitimate.

Rather it is, when we do this, the failures in rationality that we claim to find necessarily signify the presence of disturbing "external" factors. It would seem more plausible to suppose that a departure from rationality which the agent himself (or the scientific community; there are two separate possibilities) would have recongized as such, would be indicative of "disturbance," of the need for *special* socio-psychological investigation. Thus, instead of the social historian looking to the philosopher for a standard of rationality before he can begin to understand some past episode in the history of science, what would seem more likely to reveal the influence of non-epistemic factors is a historical sensitivity to the sorts of argumentation that prevailed at that time, to the special "coherence" (as Mannheim might put it) of the particular scientist's thought, and to the non-epistemic factors that would have been most likely, in that particular historical context, to interfere with the normal processes of reasoning.

And not just to interfere with, but also to *supplement*. For, as we shall see more fully in the next section, the influence of non-epistemic factors must not be simply construed as "interference." The crucial feature of science from which all discussion of these issues must begin is that theory is underdetermined by evidence (i.e. by epistemic factors), and therefore other sorts of factors are likely to have to intervene, if a scientist is to adopt one theory in the face of all the possible alternatives. Explanation in terms of socio-psychological factors is thus to be seen as a proper part of historical explanation, even in cases where epistemic factors have played the fullest role possible to them.

A second objection to the PSR can be stated in terms of Laudan's (1) above. Suppose that the historian *can* produce an adequately "rational" construal of a scientific episode in purely epistemic terms (and of course, he sometimes can), does this imply that the historical sociologist can have nothing more to add in regard to the explanation given? This seems much too sweeping, and indeed it has been strongly attacked by sociologists of science.[18] Must the apparent reasons be the "real" reasons in all cases, for instance? This is a peculiarly difficult question, as Freud and Marx and their contemporary followers keep reminding us. If the social historian can furnish us with evidence of strong social pressures or of the influence of a personality trait working to the same end as do the reasons, this must not be dismissed as irrelevant to the *historical*-causal explanation of what happened. But why admit such factors if the reasons were of themselves presumptively sufficient? Because human actions *are* so complex, and because it is often hard to know how clearly the agent perceived the structure of the argument involved. To maintain that science is "rational," self-contained, in a way that other domains are not by no means commits one to holding that scientific arguments are always so clear-cut that the influence of such factors in bringing about the victory of one or other side of a debate can be neglected.

This last is a point that the Edinburgh group of sociologists of science has ben particularly effective in underlining. According to earlier logicist views, science consists primarily of law-like assertions, and these are assessed in a relatively definitive way by means of a set of rules. The newer stress is on science as *theory,* and on the variety of criteria, none of them coercive, which govern theory appriasal. The notion of theory "acceptance" has itself been recognized to be a complex one; when a scientist "accepts" a theory, this is not normally equivalent to his claiming that it is "true." In many instances, of course, the weight of evidence may be overwhelmingly in favor of a particular theory. But the theories and models of the scientist are, of their nature, open-ended, capable of further development and of substantial modification. To go back, then, to Laudan's formulation of the PSR in (1) above, the objection is that rational appraisal is usually *not* coercive. In explaining disagreements, one can relevantly reflect on the character-traits of the researchers, for instance, without necessarily suggesting that their appraisals of their own theories is anything less than "rational."

One further shortcoming of the PSR, as enunciated by Laudan, may be noted. Let us assume that in the course of a particular episode in the history of science, only epistemic factors *did* play a significant role. Would this imply that the sociologist would have nothing to say here? Not necessarily. Durkheim reminded us that a sociological explanation can also be sought for the operation of *epistemic* factors. One could trace, for example, how the members of the group internalize the norms of the gorup. The norms themselves come to be accepted as the appropirate ones to adopt if a particular end is to be achieved. But they still have to be learnt and enforced.

Scientific rationality has taken a long while to form, and is undoubtedly not yet fully realized. It requires complex educational and communicative procedures, ones that are of considerable interest to the social historian. It is influenced by the prevailing epistemic assumptions of the society; the mode of transmission of this influence and the manner in which it makes itself felt, are worth investigation. If a scientist in the course of an investigation makes use of a simple mathematical principle (of the order of 2 + 2 = 4), one would not in writing the history of this investigation be likely to ask how this scientist came to take this principle for granted. (Of course, a psychological inquiry of the sort carried out by Piaget into how children learn such principles *could* be relevant under certain circumstances. But it would not further a standard historical account of an episode in modern physics.) On the other hand, if a scientist dismisses an hypohtesis as "*ad hoc,*" one *might* well ask: how did he learn that? What group-mechanism exists for the transmission of skills such as these? Even when a particular scientific inquiry can be laid out in straightforwardly epistemic terms, therefore, a range of sociological questions can still be put to it. Answers to the questions will often seem perfectly obvious and trivial. But to the extent that this is the case, then *this* is the objection that ought to be posed to the social historian, not the overly-strong exclusion of sociological analysis from the domain of "rational" inquiry, favored by the PSR.

5. The presumption of unrestricted sociality

The PSR, as we have just seen, involved these two claims: (1) There is a transhistorical norm of scientific rationality which can be applied to the assessment of the implicit rationality of scientific inquiry in any age. (2) The introduction of socio-psychological modes of explanation is required (and is legitimate) only when the historical episode under scrutiny fails in some way to embody the standard rationality. At the other extreme would be the view that (1) there are no norms of rationality that can claim any sort of transhistorical status; (2) socio-psychological modes of explanation apply in just the same way to "rational" processes in science as to supposed "non-rational" intrusions into it. From these as premises, depending on how, precisely, they are further specified, it is an easy step to the conclusion that since the activities of the scientist can be fully "explained" in terms of social determinants, the product of those activities can be no more than the reflection of the contingencies of the historical culture that shapes it.

The first exponents of a sociology of knowledge, like Durkheim, exempted science from the cultural relativism with which they regarded all other human products. But this exemption has been challenged by many in recent years. I mean to focus here on only one such challenge, that emanating from the "strong program"[19] in the sociology of science, promoted by a group of Edinburgh sociologists.[20] Premise (2) above is one of the cornerstones of this program, and is described by Bloor as their "symmetry principle." The sociology of scientific knowledge should, he says, be "symmetrical in its style of explanation. The same types of cause would explain, say, true and false beliefs."[21] From the standpoint of the sociologist-historian, purportedly rational pieces of scientific work ought be ascribed no sort of privilege or autonomy. They are to be causally explained, not in terms of reasons (for this would violate the symmetry principle, since some occurrences in the history of science cannot adequately be explained in terms of reasons) but ultimately in terms of non-epistemic factors of a social or a psychological kind.[22]

Bloor contrasts this "causal" model of knowledge-generation with the classical one, which he describes as "teleological" because it defines knowledge in terms of "goals" (truth, rationality) which man is supposed to have the natural capacity to attain. Such an approach relinquishes, he says, a "thorough-going causal orientation,"[23] because it explains the "rational" aspects of scientific work in an immanent way by reference, for example, to the relations between evidence and hypothesis, without involving social factors in any way. This is to detach the activity from its social ground, to make it float above the interplay of causal interactions. Bloor concludes that the teleological and the causal models "represent programmatic alternatives which quite exclude one another. Indeed, they are two opposed metaphysical standpoints."[24]

I shall call this the "principle of unrestricted sociality" (PUS). It is more extreme than PSR, which does allow *some* independent scope to social forms of explanation. As Bloor states it, PUS does not allow independent scope to "rational" forms of explanation under *any* circumstances. They are entirely reducible to the social forms which underlie (and thus account for) them. A more moderate claim would be that for a *complete* explanation of a historical episode in science, one should at some point have recourse to the social modes of explanation which would explain how the particular rational forms employed by the agent came to be adopted by him. Thus, the two types of explanation would be seen as in some sense complementary, each supplying something the other cannot. But PUS is more extreme than this, for it maintains the *ultimacy* of the social forms of explanation. It is they, in the end, that *really* explain. The "internal" historian of science who notes that the result of a particular experiment led a scientist to modify his theory is not really explaining what happened. To do that, the historian must show, in social-causal terms, how the scientist came to employ strategies of this sort in the first place. And if he can do this successfully, he can finally dispense with the "rational" modes as redundant.

The consequences of PUS for traditional views on objectivity and realism are clearly far-reaching. Bloor argues that to the extent that a theory goes beyond the empirical data, the excess (which by hypothesis, cannot be validated by any independent access to the physical reality) can only be explained by invoking on the one hand the "internal coherence" of the theory and on the other the social factors that led to its construction. In short, the "theoretical component is precisely the social component." The theory as such has no objective standing independent of the historical contingencies of the society, which entirely account for its formulation. The only way to salvage some kind of "objectivity" would be "to equate it with the social," that is, with "institutionalized belief."[25] This, of course, is not at all what has been meant by 'objectivity' in earlier accounts of science.

Barry Barnes argues that recent work in philosophy of science would justify one in holding that:

> Our current scientific models and mechanisms are likely to be seen at some future time as part of what is an

> endless chain of such mechanisms, constructed and eventually abandoned (or stripped of their ontoloigcal standing) as the activity of knowledge generation proceeds.... The diverse real universals postulated at different times and in different cultures and contexts, should be regarded alike as inventions of the mind, sustained to the extent that they are instrumentally valuable in the settings where they are found.[26]

The relevance to PUS of this sort of instrumentalist view of science is that the realist alternative would imply criteria of evaluation of a trans-cultural kind, and would challenge the *ultimacy* of social explanation in a large class of cases. Defenders of PUS have to hold that the claim that the points of light in the sky we call "stars" are enormous glowing masses of gas is not only contingent upon the kinds of society within which it was formulated but that it can be entirely *explained* by reference to non-epistemic factors operating within those social forms.

Barnes continues:

> What matters is that we recognize the *sociological* equivalence of different knowledge claims. We will doubtless continue to evaluate beliefs differentially ourselves, but such evaluations must be recognized as having no relevance to the task of sociological explanation; as a methodological principle we must not allow our evaluation of beliefs to determine what form of sociological account we put forward to explain them.[27]

The sociologist ought not, therefore, take into account the presence or absence of epistemic factors in the episode he is explaining. The form of the sociological explanation offered, and the completeness of that explanation as an explanation, in no way depends on whether the scientist studied was guided by reasons or not. Barnes does not think that this symmetry claim leads *immediately* to relativism; rather, "the naturalistic equivalence of the knowledge of different cultures" is "merely a finding, something which happens to be the case."[28] Whether immediately implied or not, however, it is clearly a very short step from PUS to a rather far-reaching relativism, a step which

Bloor and Barnes are quite willing to make.

What are we to say to this? A simple clarification first. Someone who rejects the symmetry principle is not necesarily (as Bloor supposes) relinquishing causal explanation. To say that the reason why Kepler gave up the attempt to describe the orbit of Mars in terms of circles was that he saw that Tycho's data could not be reconciled in this way is *not* to imply that his behavior was somehow uncaused. He had a reason to decide as he did, and his perception of this *as* a reason, was the cause of his decision. To explain the decision (causally), one would cite the reason, as Kepler himself did. And this would be a sufficient explanation, under most circumstances.

One could, of course, press further and ask what had led Kepler to perceive a reason of *this* kind as a sufficient one for such a decision. The answer to this could be complex, because one could respond in terms of reasons (Kepler had *reasons* to accept the principles of rationality he did) or in terms of social causes (how, for example, he came to learn these principles in the first place). But to assume that in this regress, the ultimate level must always be the social one, that this is what *really* explains, begs the question. We shall return to this in the final section.

How are "social causes" to be discovered and how do they function as explanation? A social cause is an instance of a regularity which can be specified in terms of the standard parameters of sociology. X's membership in group G is said to "cause" him to behave in manner B, if membership in G is correlated in a statistically significant way with behavior B. To "explain" here is to subsume under an observable regularity; this is standard nomothetic explanation of the sort celebrated by Hume.

Some illustrations may help. Steven Shapin has shown, for example, that support for phrenology in Scotland in the early nineteenth century correlated fairly well with membership in the urban middle class, whereas the critics of phrenology were predominantly drawn from elitist upper-class circles.[29] Here the correlation is with social class. Other recent essays have focussed on national groupings. Joan Richards shows, in an interesting paper, that British mathematicians were much more sceptical of non-Euclidean geometry in the mid-nineteenth century than were their Continental colleagues.[30] There are, of course, some well-known difficulties to overcome in arguing from a correlation of this sort to a genuine causal connection.[31]

But a more serious challenge for PUS is that nomothetic explanation is only a preliminary step. To explain the expansion of a piece of iron placed in the fire by saying that iron always expands when heated is only a first stage. The scientist immediately asks *why* the regularity itself holds: what must iron be like in order for this to occur? And it will not ordinarily be possible to answer this simply in terms of other observed regularities.[32] The situation in the social sciences is admittedly not the same in this regard as in the natural sciences. But when we have noted the correlation between support for phrenology and membership in the middle-class, we are immediately led to ask *why* this correlation should hold. Was there something about the *content* of phrenology that made it more acceptable to middle-class people? But this brings us right back to epistemic factors again. Social groups may also be marked off (as we saw in an earlier section) by their beliefs. Cantor notes, for example, that the protagonists in the phrenology debate were separated by "major differences in philosophical, metaphysical and methodological principles, and also in specific cases, in theological out-look."[33] How is one to decide whether to appeal ultimately to differences of social class or to differences of belief in explaining the course and outcome of this debate? Bloor, of course, assumes that the social factors are the ultimate in explanation. And in some cases they are, no doubt. But even in cases where they are, how exactly would this be *shown?*

How does a "social factor" operate as a cause to bring about the holding of a particular belief? There is no single answer to this. Class-membership can operate as a non-epistemic factor influencing a person to one side rather than the other in a complex disagreement of belief. But even in such cases, reasons will ordinarily play a part too. Why did X reject phrenology? Was it because he had specific epistemic objections to it? Or was it because of his social interests as a member of the urban middle-class? Or both? This is where the sensitivity of the historian is needed. A decision cannot be made in such cases on the basis of an *a priori* methodological principle which asserts that explanation in terms of social factors must be regarded as equally helpful in all cases.

It is no accident that historical sociologists focus on *controversies* in past science. If a strong correlation can be found between social class, nationality, or the like, and the side a person took in the controversy, it gives a *prima facie* case for thinking that explanation of the course of the debate ought rely, in part at least, on social analysis. But what of the remainder? The normal process of science involves a fair amount of disagreement about procedures, reliability of results, validity of hypotheses, and so on. But *controversy,* meaning: a public and persistent contention between two or more well-defined "sides", is less common.[34]

In explaining a historical episode where no controversy occurred, the historical sociologist is less likely to find significant correlations, since in order to do so he must link the presence of one or more social parameters with the use of a particular principle or the defense of a particular position. If those involved in the episode exhibited no sharp differences, the best the sociologist can do is to ask whether their common views could be socially accounted for by some factors common to the society as a whole. But this is less obvious as an explanatory strategy; the more obvious solution in such cases will simply appeal to scientific "reasons."

But even where controversy *did* occur, it by no means follows that the various sides must neatly correlate with explanatory social factors. The fact that theory-choice is generally underdetermined does not (as Bloor supposes) imply that the requisite extra determination must come from social factors. If X and Y attribute different relative weights to the criteria of predictive accuracy and of fertility, this may lead them to choose differently among theoretical alternatives. But their differences of perception in regard to the criteria are not necessarily traceable back to some sociologically definable difference in the groups to which each scientist belongs. Once again, each may have *reasons* for the preference he exhibits. To say that such reasons *must* have further socio-psychological determinants simply lays down an *a priori*.

Kuhn is perhaps the person who has most effectively argued that science has to be considered a social product. Yet he went on to maintain that over the course of time the practitioners of the more mature sciences have developed techniques and criteria that enable them to solve the puzzles their sciences constantly throw up:

> As a result, they constitute a special subculture, one whose members are the exclusive audience for, and judges of, each other's work. The problems on which such specialists work are no longer presented by the external society but by an internal challenge to increase the scope and precision of the fit between existing theory and nature.... In short, compared with other professional and creative pursuits, the practitioners of a mature science are effectively insulated from the cultural milieu in which they live their extraprofessional lives. That quite special, though still incomplete, insulation is the presumptive reason why the internal approach to the history of science has seemed so nearly successful.[35]

Where Kuhn sees insulation, Bloor sees total interpenetration. Indeed following Mary Douglas, he argues that the scientist's perception of what is right and necessary in social relations tends to generate the causal relations he is going to find in nature.[36] But is this really the case in organic chemistry, in high-energy physics, in molecular biology? Kuhn may have overstated the degree to which these sciences are insulated (although he is careful to note that the insulation he speaks of pertains only to concept and to problem-structure). But he is surely right that the social factors that bear on controversies in these sciences nearly all originate within the scientific community itself. Though sociological techniques can help us understand better the course of the debates in high-energy physics over recent decades,[37] they give us no reason to suspect that defenders of the S-matrix theory share a social classification which would explain why they prefer S-matrices to quarks.

Sociological analysis, in cases like this, may allow us to understand how the groups formed, how they came to reflect personal or institutional interests, and so on. But it will not ordinarily explain the *content* of the theory defended: why *this* group should defend S-matrices, rather than some other alternative. (Note the difference between this case and the phrenology one, where there was, supposedly, a reason why one particular socially-definable group should support a theory with a particular content.) We are thus forced to have recourse also, if we want to understand the course of the scientific debate, to an account in terms of epistemic factors.

Social factors have quite obviously influenced the course of many controversies in the history of science. Defenders of PUS draw upon a number of favorite cases, which serve a purpose similar to that of the oft-told tales of caloric and phlogiston in the repertoire of anti-realists. There is the Darwinian controversy, of course, and the later eugenics debate. A recent favorite is the Pasteur-Pouchet debate, no longer considered a model of scientific rationality on the Pasteur side. These topics, it ought to be noted, form a rather special class. They nearly all bear on the status of man or on the relations of man and society. Can one properly make an inductive generalization from such cases as these so as to make PUS equally applicable to *all* contexts in the natural sciences, organic chemistry as well as evolutionary biology? The sample so far has been a biased one, not at all adequate to the weight now being laid on it.

The objection to PUS is not that sociological analysis is unable to illuminate episodes in the history of science of the most respectably rational kind. Of course it can. The issues are, first, whether one can in all cases explain the *contents* of scientific theories in this way, and second, whether the ultimacy of such explanation is a matter of *principle*. The "symmetry principle" embodied in PUS seems to require an affirmative answer to both questions. Anything less would allow an explanation in terms of reasons to take priority in most cases.

One further difficulty with PUS must be mentioned. Bloor lays down as one of the four theses of the "strong program" that sociology of science should be *reflexive,* that is, that "its patterns of explanation would have to be applicable to sociology itself."[38] He does not pursue this theme further, but its implications are potentially destructive. To "explain" the "strong program" one would then presumably have to look to the social interests of the group that created it, to their identification with anti-science elements in the broader society perhaps, to their concern as a professional gorup with the limitations imposed on their scope by philosophers and internalist historians, to the rewards of notoriety accruing to simply-stated extreme formulations, to the excitement of challenging the established sources of power, and so on. So far as I know, no member of the group has as yet provided us with the requisite analysis, and I am certainly not competent to do so myself.

It is worth asking, however, just what the outcome of such an analysis would be. Could it serve as criticism or justification of the program itself? It will depend on whether the distinction between science and ideology can be drawn. If it cannot, as Bloor and many other recent sociologist-historians of science have suggested, then there is no reason for anyone who does not share the social intersts of the group to accept the program itself. Unlike the natural sciences, there is no question here of prediction and control. What we have is a theory of knowledge, a wide-ranging philosophy of social science which proposes regulative principles for which no general basis in historiographical practice is given. Bloor's own objection to a philosophy of this sort is that it is no more than the "affirmations of the values and perspectives of some social group,"[39] thus making effective criticism impossible. How is he to avoid this charge himself? If he uses *argument* in his defense as he does in his book, can one not, according to his own principles, explain what is going on at a deeper level by the social interests from which the arguments flow? And since these intersts themselves cannot, without circularity, be defended but can only be referred to the society that embodies them, there seems to be no escape from scepticism.

6. A certain autonomy of the rational

A historical explanation for the support a scientist gave to a particular theory will ordinarily invoke the relation of evidence and hypothesis, and appeal (implicitly, at least) to some sort of norm of rationality in doing so. The historian may choose to push further and ask whether there were social factors influencing the choice of this particular theory. He may also inquire whether the norm of rationality itself had some sort of social forces operating in its favor, serving not as warrant but as non-epistemic causal agency. But what he will assuredly *not* maintain *a priori* is that a sociological or psychological investigation will be bound in all cases to illuminate why theory A was adopted over theory B. To explain this latter, the connections of *this* theory and *this* evidence will as a rule have to be worked out, and their adequacy discussed in terms of someone's canons of rationality. The epistemic factors cited here may well be non-standard, as when historians cite neo-Platonic metaphysics in discussing some of Kepler's views. But they are still *epistemic,* a point that we have been stressing throughout this essay.

There is, however, a much more important point to be made. Is it the case that the rational norms of science have no sort of transcultural standing that could exempt them from further explanation in terms of the culture in which they are employed? There are several possible ways of defending an affirmative answer to this all-important question. One is to invoke a "natural" rationality, a human ability to follow correct reasoning. This is the sort of move that Bloor and Barnes strenuously oppose. Bloor argues that even mathematical reasoning, "pure" though it may seem is not exempt in this way.[40]

A more immediately effective response is suggested by the reference both Bloor and Barnes make to a social interest in prediction and control which must be allowed as a defining characteristic of science. Barnes says:

> Knowledge cannot be understood as more than the product of men operating in terms of an interest in prediction and control, shaped and particularized by the specifics of their situations.[41]

The question, of course, will be whether the goals of prediction and control will not end by making knowledge considerably more than *just* prediction and control. After claiming that "scientific theories, methods, and acceptable results are social conventions,"[42] Bloor defends himself against the charge that this would be to make science an arbitrary affair. Conventions, he responds, are not arbitrary; they have to work within the limits of social credibility and practical utility. This latter can be specified in terms of the criterion of successful prediction and control. "This is a harsh discipline to impose on our mental constitution, but it is no less a convention."[43] Conventions, he reminds us, can be quite demanding; recall, for example, the feats of endurance demanded of North American Indian males in order for them to be initiated as members of the tribe.

There is, however, a crucial difference between the two cases. The Indian custom was culture-bound; there was nothing about tribe-membership which implied that *these* feats of endurance were the appropriate test. In other cultures, quite different requirements were found. Bloor does not show that the same is true of the prediction criterion. Might there be cultures where science would *not* be held to it? Would we describe as "science" an activity to which successful prediction was irrelevant? The answer is surely no to both questions. An activity which did not qualify at all under this criterion simply does not qualify as "science" either. Nor is this a chauvinistic modern-mindedness. It is a matter, in the first instance, of getting our use of terms right.

But, of course, the matter is much more complicated that this. Science is not just a matter of accurate prediction. It is also a search for *understanding,* and it could be argued that this, in human terms, is even more basic than predictive accuracy. Further, the norms of prediction and control are by no means equivalent, and have not had the same history. So an intricate story of how these norms have come to be successfully interlaced would have to be told. The historical notion of science is not sharply delimited; there have always been difficult boundary-decisions to make.

Nonetheless, there is a high degree of continuity in the story. And the goals of prediction and control may be as old as man himself. Thus there *is* a "teleological" account, to use Bloor's term, which recognizes a gradual historical development from an original very general desire for reliable knowledge of the environment, and for an understanding of puzzling events, to the complex of procedures and norms that go under the name of "scientific rationality" today. There has been a constant *learning* process, a discovery of what will work, as one criterion is tested against another.

The growing reliance on theoretical entities in the natural sciences after 1700 A.D. was due in the first instance to their indispensable role in prediction, but later to a growing conviction also that (outside the problematic area of mechanics, at least) a world of real structures was gradually being disclosed, a world of chromosomes and binary stars and a myriad other differentiations. The spiral structure of the DNA molecule is not in any interesting sense a product of mid-twentieth-century British culture; it is a product rather of a centuries-long effort spanning many cultures of the most widely diverse sort. What is common to this long, and still unfinished, development is the human desire to know.

Social factors have played a central role in this development. But what must be insisted on is that scientific rationality has attained a degree of autonomy in consequence of the successful testing of human constructs that has gone on in a systematic way now so long. It is not the case that an element in it could be challenged successfully on social grounds; the Lysenko case showed that. We are not at liberty to modify *except* in response to the internal epistemic demands of science itself.

What is particularly troubling about PUS is the implication that science and ideology cannot be separated. Not everyone draws this implication. Barnes, for example, distinguishes between legitimate epistemic interests and social intersts that can potentially defeat the epistemic ones. These latter are, in his eyes, illegitimate and constitute "ideology."[44] There are ways, he reminds us, in which ideology can be detected and separated off. But this distinction would seem to violate the symmetry principle, since legitimate and ideological factors are being separated on the grounds of their origin. Though both may be said to proceed from "interests," what constitutes (as far as I can see) the legitimacy of one set, by contrast with the illegitimacy of the other, is not just derivative from the society in which science is carried on, but from the entire history of science, as a *cognitive* enterprise of a particular sort.

Bloor, more consistently perhaps, does not draw this distinction:

> All knowledge, the sociologist could say, is conjectural and theoretical. Nothing is absolute and final. Therefore

> all knowledge is relative to the local situation of the thinkers who produce it.[45]

It is the "therefore" here to which we have been objecting. From the fallibilist view of sciences as conjectural and theoretical, one simply cannot infer to the notion of objectivity as a "social phenomenon" which Bloor tries to enjoin on us. It is, in the end, the sense in which knowledge is *not* "relative to the local situation of the thinkers who produce it" that historians of science have most often found profitable to pursue in their work.

Somewhere between the extremes of PSR and PUS, the working historian will continue to ply his trade, taking each case as it comes and ignoring the prescriptions of philosophers and sociologists. Why did Newton so obstinately defend the principle of the passivity of matter, despite the fact that his theory of attraction seemed to give an overwhelming reason to treat matter as active? Was it because of a theology and a metaphysics which attributed to God total dominion over material things?[46] Or was it because "behind it must lie ... an interest and a social purpose" in legitimizing the hierarchical social structures in the England of the day?[47] Or might both explanations be correct? This sort of investigation will not be helped by an appeal to universal presumptions. Rather, what is needed is a detailed and historically sensitive analysis of reasons and motives, in which the status of neither the rational nor the social is ever taken for granted.

Department of Philosophy
University of Notre Dame

Notes

* The germ of this paper may be found in an earlier one, "The rational and the social," *Grazer Philosophische Studien*, 12/13, 1981, 13-33. I am indebted to Alasdair MacIntyre and Larry Simon for their comments on that earlier paper, and also to the participants in the three-year study group organized by the Hastings Center, and directed by Art Caplan, investigating the topic of "The closure of controversy in science and ethics." Some of the ideas developed below were first presented to this

group, and benefited greatly from the discussion they received there.

1. See G. Holton, *Thematic Origins of Scientific Thought: Kepler to Einstein,* Cambridge (Mass.): Harvard University Press, 1973, pp. 219-259.
2. New York: Macmillan, 1934.
3. "Weimar culture, causality, and quantum theory, 1918-1927: Adaptation by German physicists and mathematicians to a hostile intellectual environment," *Historical Studies in the Physical Sciences,* 3, 1971, 1-115.
4. New York: Columbia University Press, 1981.
5. *Op. cit.*, p. 372.
6. For a fuller discussion of Graham's book, see McMullin, "Values and the understanding of science," *Hastings Center Report,* in the press.
7. For a discussion of this example, see M. Hesse, "Hermeticism and historiography: An apology for the internal history of science," *Minnesota Studies in the Philosophy of Science,* volume 5, 1970, 134-159.
8. For a detailed elaboration of Newton's *Principia* along these lines, see McMullin, "The significance of Newton's Principia for epistemology," in *The Epistemology of Science: The Views of Four Great Physicists,* ed. J. Blackmore, in the press.
9. "History of science and its rational reconstructions," *Boston Studies in the Philosophy of Science,* volume 8, 1971, 91-136.
10. McMullin, "The history and philosophy of science: A taxonomy," *Minnesota Studies in the Philosophy of Science,* volume 5, 1970, pp. 32-34.
11. I. Lakatos and E. Zahar, "Why did Copernicus' research programme supersede Ptolemy's?." in *The Copernican Achievement,* ed. R. Westman, Berkeley: University of California Press, 1976, 168-192.
12. G. Currie, "The role of normative assumptions in historical explanation," *Philosophy of Science,* 47, 1980, 456-473. Currie too readily concludes that "normative

assumptions" (an imputed rationality) cannot under *any* circumstances be employed when "explaining" episodes from the history of science. He does not allow for the possibility that there may be a second mode of "explaining" which does *not* purport to tell us why the agents acted in the way they did. This leads him to exclude *any* form of "reconstruction" of case-studies in indirect support of a (contemporary) theory of science. But this goes far beyond any arguments he appears to offer.

13. Berkeley: University of California Press, 1977.
14. In more recent essays, such as the one contained in this collection, "The pseudo-science of science," he has to my mind retreated from the more extreme statement of a PSR in the 1977 book. We shall focus here on the position he defends in the book, in order to contrast it with the extreme position defended by some advocates of the "strong program."
15. *Op. cit.*, pp. 208-9.
16. *Op. cit.*, pp. 128-133.
17. *Op. cit.*, p. 209.
18. For a particularly caustic assessment, see B. Barnes, "Vicissitudes of belief," *Social Studies of Science*, 9, 1979, 247-263.
19. I prefer this to the label 'strong thesis' proposed by M. Hesse (*Revolutions and Reconstructions in the Philosophy of Science*, Bloomington: Indiana University Press, 1980), because of the play on two incompatible senses of the term 'strong' that (as many have noted) make the label 'strong program' a little too self-congratulatory.
20. Those most closely asssociated with it are David Bloor, Barry Barnes, Steven Shapin, and Michael Mulkay. The term 'social historian' might just as well be applied to them. Though members of the group do employ sociological techniques on occasion, it is with a view to describing and explaining historical episodes in science. The boundary between historiograplhy and sociology is notoriously ill-defined. But the Edinburgh group, to my mind, is situated well to the side of history. And some of them (Barnes and Bloor, notably) devote much of their

attention to a second-level philosophical account of their field. So the term 'sociologist' is inadequate to convey the range of their interests.

21. *Knowledge and Social Imagery,* London: Routledge, 1976, p. 5.
22. Edward Manier reminds us that there are some important differences within the Edinburgh group, particularly in regard to the degree to which ideals of objectivity and rationality are culture-dependent. See "Levels of reflexivity: Unnoted differences within the 'strong programme' in the sociology of knowledge," *PSA 1980,* E. Lansing: Philosophy of Science Association, vol. 1, 197-207.
23. *Op. cit.*, p. 8.
24. *Op. cit.*, p. 9.
25. *Op. cit.*, pp. 86-7.
26. *Interests and the Growth of Knowledge,* London: Routledge, 1977, p. 24.
27. *Op. cit.*, p. 25.
28. *Loc. cit.*
29. S. Shapin, "Phrenological knowledge and the social structure of early nineteenth-century Edinburgh," *Annals of Science,* 32, 1975, 219-243; "Homo phrenologicus: Anthropological perspectives on an historical problem" in *Natural Order: Historical Studies of Scientific Culture,* ed. B. Barnes and S. Shapin, London: Sage, 1979. This latter collection of essays is designed to illustrate the manner in which social realities shape cognitive forms in science.
30. "The reception of a mathematical theory," in *Natural Order,* pp. 143-166.
31. See the critique of Shapin's thesis by G.N. Cantor, "A critique of Shapin's social interpretation of the Edinburgh phrenology debate," *Annals of Science,* 33, 1975, 245-256.
32. See McMullin, "Structural explanation," *American Philosophical Quarterly,* 15, 1978, 139-147.
33. "Phrenology in nineteenth-century Edinburgh: An historiographic discussion," *Annals of Science,* 33, 1975, 195-218; p. 217.
34. See McMullin, "How controversies end in science," in The

Closure of Controversy in Science and Ethics, ed. A. Caplan, to appear.

35. *The Essential Tension,* Chicago: University of Chicago Press, 1977, p. 119.
36. "Durkheim and Mauss revisited: Classification of the sociology of knowledge," in *The Language of Sociology,* ed. J. Law, 1981, sec. IX.
37. See Diana Crane, "An exploratory study of Kuhnian paradigms in theoretical high-energy physics," *Social Studies of Science,* 10, 1980, 23-54.
38. *Knowledge and Social Imagery,* p. 5.
39. *Op. cit.,* p. 71.
40. For a critique of this part of Bloor's argument, see Gad Freudenthal, "How strong is Dr. Bloor's 'strong programme'?", *Studies in the History and Philosophy of Science,* 10, 1979, 67-83.
41. *Interests and the Growth of Knowledge,* p. 24.
42. *Knowledge and Social Imagery,* p. 37.
43. *Op. cit.,* p. 39.
44. See the discussion in Manier, *op. cit.,* p. 201.
45. *Op. cit.,* p. 142.
46. As I argue in *Newton on matter and Activity,* Notre Dame: University of Notre Dame Press, 1978.
47. Bloor ("Durkheim and Mauss revisited," Sec. XI), urging the thesis of M.C. Jacob's book, *The Newtonians and the English Revolution,* 1689-1720, Ithaca: Cornell University Press, 1976.

Ian Jarvie

A PLAGUE ON BOTH YOUR HOUSES

The debate between Larry Laudan and David Bloor is about the limits of the sociology of science. Hitherto the content of science, its ideas, had been declared out of bounds by no less authorities than Scheler and Mannheim,[1] but the Edinburgh-based advocates of the Strong Programme are trying to run the blockade. All science is ideology, they say, not least the sociological science of science-as-ideology. Laudan put a shot across Edinburgh's bows with his argument that most of their claims are programmatic; he might have added the torpedo that few of the cases they cite concern scientific ideas; rather do they look at the *uses,* or *reception* of, or *priority disputes* about, ideas. Ideas in their social dimension are not the same as ideas as such. But Laudan overlooks this decisive point, as does Bloor, with disastrous consequences, as we shall see. Instead, in his book *Progress and Its Problems* Laudan tried to blockade science against the sociology of knowledge by conceding that scientific ideas are sometimes legitimately susceptible of sociological explanation. The concession was nugatory: sociology was admitted when all else failed. Not surprisingly, Bloor vigorously attacked, trying to vault the blockade and show his troops were already inside Laudan's exclusive zone and that, moreover, sociological explanations were better tested case by case in the scientific manner, than by argumentative closure. Laudan set strict limits to social explanation in science; Bloor set no limits in principle. By contrast, the thesis I shall offer in this paper is that there can in principle be no sociological explanations of science since, while there are, it is true, sociological aspects to it, science as such is other quite different things which have to be discussed in a completely non-sociological way.

The debate between Laudan and Bloor conceals a deep agreement on assumptions I want to contest. Two minor

J. R. Brown (ed.), Scientific Rationality: The Sociological Turn, 165–182.

examples are Laudan's ready agreement to dismiss, with Bloor, all transcendental considerations (Laudan 1981, p. 184). A second is Bloor's airy confession that he is an inductivist (Bloor 1981, p. 206). While Laudan expresses some incredulity at this, Bloor's counterclaim that Laudan is a deductivist is false. Laudan too, as I argued elsewhere (Jarvie 1979, pp. 494-6), is an inductivist. The major issues over which I wish to contest their agreement are three: 1) the attempt to demarcate internal from external considerations in science; 2) the attempt to demarcate rational from irrational beliefs; 3) the attempt to discuss science in terms of scientific beliefs. My theses shall be 1') that the demarcation between internal and external is a metatheory of a discipline and not a natural boundary observable in the phenomena; 2') that beliefs are neither rational nor irrational; 3') that science is a system of statements, not beliefs, and, because statements are not merely entities in the social world, social explanations are in a strict sense inadequate to explain important aspects of science, namely ideas.

1. The Internal and the External

It was the late Imre Lakatos who most sharply wielded this demarcation. He divided the progress of mathematics into its rationally reconstructed part, which took the form of a critical and progressive debate according to the logic of conjecture and refutation, and its actual history, in which the internal logic of the debate might not be followed at all (Lakatos 1976). To the charge that his rational reconstruction was merely a caricature of the real history he replied that one could just as well argue the reverse. His own sympathies seemed to incline towards regarding rational reconstructions of debates in mathematics and science as rational, while the mundane contingencies of what actually happened were irrational to the extent that they deviated from the rational. Such deviations were consequent on human error, psychological quirk, social 'interests', and the like. These were 'external' to the world of ideas, hence his implicit equations: rational = internal, irrational = external.

The logic of progress in rational/internal science was the Socratic logic of problem, conjectured solution, criticism, attempted refutation, new solution, and so on. Thus approximation to critical debate became a norm of rationality and was called the internal part of science, all the rest was external. When Popper declared that he know of nothing more rational than a well-conducted critical discussion (1972, p. 22) I doubt he intended 'well-conducted' to be equated with a kind of inner logic that could have normative force. Provided immunising strategies are not employed, the possibility of there being quite a few well-conducted critical discussions about a single problem seems obvious. After all, formulating the problem, creating solutions and dreaming up arguments for and against all involve creative acts. If there were an inner logic to creativity it should be easier and more predictable (Jarvie 1981).

Lakatos' use of Hegelian language and jokes might be interpreted as a diversionary tactic to avoid being pressed too strongly on whether there 'really' is an interior and exterior to science. My guess is that, if cornered, he would readily have conceded that nothing is hard and fast and such distinctions as internal/external are created for a purpose. For another purpose they might not be useful. What was Lakatos' purpose? An objective answer would be the problem-situation created in Popper's philosophy by Kuhn's criticism of it. Popper hailed Kuhn as his best critic (1970) and perhaps thereby stimulated his colleagues to take more notice than they otherwise would have of Kuhn. Kuhn argued that in truth scientists were dogmatists and consensualists, not open-minded fallibilists (Kuhn 1962, 1970). Science began when critical discussion (on fundamentals) ceased. Lakatos' solution is very ingenious: accept both Popper and Kuhn but restrict them to different domains. Popper has described the ideal logic of rationality recontructed science, Kuhn has described the messy practices of actual scientists. This creates for Lakatos the further problem of the relationship between science in the flesh and science in spirit. Lakatos' solution to this problem is his epicyclical development of research programmes. Research programmes have hard cores that can be dogmiatically defended and periphral areas where critical discussion is freely indulged. The historian and philosopher of science can thus mediate between the ideal world and the real world.

Taking up where Lakatos left off, Laudan invests the internal/external distinction with a great deal of significance. But it does not quite coincide for him with the rational/irrational distinction. Science consists partly of beliefs caused by reasons, and partly by beliefs we cannot give reasons for. Hence laudan concedes Bloor's point that sociological and other seemingly external kinds of matters can penetrate the heart of the internalist citadel. In his 'More on Bloor' he adds a new metaphor with intriguing consequences for the internal/external debate. He compares internal explanation of the progress of science to explanations of the 'routine working of a mechanical clock entirely within the framework of the science of mechanics, without talking about chemical processes of oxidation (1982, p. 73). But when a clock malfunctions because of rust then those different causal factors have to be appealed to in explanation. This is a rich metaphor, especially as Laudan allows that it precisely corresponds to the differences between rational and irrational beliefs (of which more below). It could for example be extended a bit: we might trace the malfunctioning of the clock to rust brought on by a natural factor like humidity, or we may trace it back to a human factor (social? psychological?) like neglecting to clean and lubricate it periodically.

What line does the clock metaphor help us to draw between the internal and the external aspects of science? The answer would seem to be goal-directedness. When the clock is achieving its goal of routinely telling the correct time, then the principles of mechanics and metallurgy employed in its manufacture suffice to explain it. When it malfunctions we could wonder if it was refuting those theories, or we could guess that factors we had minimised and discounted, such as increased friction due to rust, were coming into play. When thought progresses smoothly, solving problems as it goes, then rational factors like arguments and reasons are sufficient to explain it. But when things go wrong, when thought malfunctions, and science goes into blind alleys, embraces false theories, becomes prone to superstition and metaphysics, then we need to bring in interfering factors.

Bloor should oppose this neat metaphor because of his symmetry principle, namely that true and false ideas, functioning and malfunctioning machines, should be explained in an even-handed way. Yet I suspect he does not conceive of the inductivist mind moving along from success to success unless deflected by interfering factors. His remarks (1981) suggest that he would place very little credence in Laudan's notion of a routinely and hence successfully functioning machine. We have already seen that it is his view that there are always social, and economic and psychological factors at work in the world of science, stimulating interest in problems, galvanizing people towards solutions, and adding to the pressures of debate between rivals. His examples are: a priority dispute between Newton and Leibniz; the politics of the rejection of causality in Germany; social interests in statistical theory; politics and phrenology; the politics of a revolution in physics; professional and external factors in the race-intelligence dispute; German universities and research; and the acceptance of iatromechanism by the College of Physicians.

These examples certainly make his case that he does not restrict scientific sociological examination of science to malfunctioning cases. He sets no *a priori* limit but rather looks at each case as it comes along. He is, however, a little carried away. For example he writes of these highly contentious case studies that parts of science have 'demonstrably shown themselves affected by the presence of social interests' (200). The demonstrability is, I fear, mainly in his own mind and that of his colleagues and friends. Again, a

> general historical picture is now emerging of this phase of our scientific development (200).

Note how the truth is at last emerging, how highly arguable interpretations are treated as facts. I fear the only facts in the list of writings Bloor cites are the bibliographic specifications. It is I think notable how often his examples are about *priority, reception* and *acceptance,* all matters one might consign to the social history of science and legitimately leave out of science itself.

While I do not want to side with either Bloor or Laudan in this dispute, but rather to undermine both their positions, I incline a little more to Laudan than to Bloor on one central issue, namely that science is unique among intellectual and social achievements and must be treated as such. Where Laudan makes a wrong move and invites Bloor's attack is in his mechanical model of routinely functioning rationality. This allows Bloor to trundle in Barnes' 'natural rationality' and accuse Laudan of equivocating between descriptive and normative (see below). The first point to note is that a problem, solutions to that problem, tests of those solutions, and a best guess among the rivals all have those *intellectual* statuses whether they have the same social statuses or not. No matter whether the scientific community cannot see the problem, or refuses to accept the solution. That does not make the problem any less of a problem or the best solution any less a solution or the best solution. Phlogiston is part of the history of science, both as a solution to certain problems and as a doctrine that was eventually overthrown. That the chemical community now tries to pretend that phlogiston was some kind of perverse going wrong, that the chemical community at the time resisted what became known as oxygen, is incidental information, no more. Priestley's ingenious arguments, like Berkeley's against Newton, are an indication that there is no 'natural' or 'routine' inner logic to the actual progress of science. It is a human creation subject to the concomitant categories and we can but attend to its fitful movement and not apply the hindsight of assuming we should see steady progress towards what is presently accepted as correct. So on this matter I side with Bloor in arguing that science should not be divided into routinely functioning and malfunctioning, with different types of explanation mandated for each. Rather must science be allowed to have a real history of stops, starts, side-tracks, pseudo-problems and struggle. All of these categories, though, are intellectual categories, not social, political, economic or psychological ones.

Something needs to be said to correct the misleading picture of the social dimensions of science adopted by both Laudan and Bloor, and to defuse their dispute about internal versus external factors. Social institutions are abstractions that we constitute out of patterns of social interaction. The interplay of expectation and action, as I have explained elsewhere (1972), is quite subtle. Hence, science, as a social-institution among others, is an abstraction from such things as: science lessons in school, people in white coats in laboratories, shelves of books and journals in libaries, conferences like the AAAS, historical traditions of endeavour, the invisible college, university departments and degrees, Nobel prizewinners, expert witnesses in court, medical researchers, and so on. These are the common-or-garden elements from which the general public builds up an imagine of science. To even conceive of this highly abstract social institution as having an inside and an outside strikes me as ludicrous. Further to image that there is some inner logic or routine functioning that marks it off from outside social pressures is even more so. Science in its actual social embodiment is enmeshed with its wider society in countless ways.

Yet when we do abstract from these concrete manifestations and talk of 'science' have we not a somewhat narrower and purer notion of science in mind? Is it foolish to be less cynical about science than about journalism, politics or industry? I do not think so. Science is more and better than the sum of its social parts; to think otherwise is reductionism. Despite its human lack of perfection, science is an example of a uniquely successful social institution created to inquire into the way the world is. Many other social institutions have undertaken that aim, including religious schools, governments, and other interested parties. None of these ever succeeded in achieving the disinterested detachment from social interests that science did. Some four hundred years ago there began a revolution in thought that marked a decisive break with the past. Only this claim unites and gives signficance to the various components I listed above as constituting the actual social embodiment of science. Were science not something new and special then we could see it as just another religion, or political party, or means of ideological manipulation. Sometimes I wonder whether the advocates of the strong programme are into such a debunking mode towards science. In my review of Laudan's book (1979) I argued that the great damage Kuhn had done was to offer a way for jealous sociologists to demystify natural science and to not so subtly debunk it in the process. My own paltry attempt to re-enchant it has evoked reactions that suggest my diagnosis was correct.[2]

Ernest Gellner has argued that the rise of science and of the industrialization based upon it marks a sea-change in the social history of mankind, in the world and what we make of it (Gellner 1964, 1975). But he avoids the usual relativism by arguing that not only is industrial society more affluent and powerful than other societies but that its citizens are especially privileged because their social order is better (Gellner 1964, pp. 33-40) and has, collectively, more knowledge about the world and the way it works. Gellner calls this the growth of cognitive power. Cognitive power is power based on true or correct understanding of the world and hence the most effective means of goal-achievement. Mankind has a strong practical/social interest in increasing this power but, curiously, it is only by detaching the quest for knowledge from such 'interests' that it expands cognitive power. There is then no 'natural' or given demarcation between the internal or the external, but rather there is the aim of the disinterested pursuit of knowledge which acts as a regulative idea and forces us to recognize that failure to understand, adopt, accept or exploit a scientific discovery does not make it any less a discovery in the objective, disinterested intellectual sense; acceptance and belief in a pseudo-discovery does not make it any less pseudo. The old-fashioned name for cognitive power was truth, the old-fashioned label for the philosophy that says science tries to find out the truth of the way the world is is realism.

Gellner in a forthcoming book agrees that realism makes sense of the way science proliferates further problems, but he declares himself allergic to emphatic assertions of realism precisely because it does not appear to be testable.[3] It may not be globally testable, and in that sense is a metaphysics, but broken down into scientific components it is constantly tested when scientists try to find out if they have a result or an artifact. Yet in his *Times Literary Supplement* review of Barnes, Gellner (1982) firmly repudiates Hesse's 'Finitism' -- an extreme form of nominalism. Rather than embracing Platonism he argues for the importance of the difference between societies that are blithely finitistic and those which at least strive to free congition from social interests.

Since Laudan too is on record as opposed to realism, here is perhaps the place to offer a simple argument for it. Whatever its difficulties, realism is the simplest hypothesis that accounts for the coordinated success of science. By the coordination of science I mean not just between separate research groups and different countries, but also between the scientist, his problem and his tests. That the scientist can coherently engage in all the things he does, and communicate about it, and achieve something we readily recognize as progress, makes the best sense if we assume that what is happening is that somehow these efforts are putting us into contact with a single, 'real', world. Our predicament may strongly resemble the blind men trying to describe the elephant, but the difficulty of the task, the necessary component of metaphysics, is not a reason for us to concede that the elephant is either an illusion or a convention.

2. The Rational and the Irrational

Laudan's arationality assumption says that '*the sociology of knowledge may step in to explain beliefs if and only if these beliefs cannot be explained in terms of their rational merits*' (1977, p. 202, italics in original). Some beliefs, then, have rational merits (reasons) and others do not. Bloor goes along with this 'very elementary representation of man's natural rationality' (207). 'For my part I am as happy to see the word "rationality" used to refer to a set of standards as to see it used to refer to man's innate cognitive competences' (209). But he states that 'all institutionalised systems of belief are compatible with plausible models of natural rationality' (208-9). So an 'asociological model of good reasons' mixes natural rationality and normative rationality.

In the previous section I criticised Laudan's idea that there is a clockwork or algorithmic functioning of the mind that results in beliefs for which there are good reasons and hence no further explanation is required. To the extent that Bloor has argued social considerations cannot be excluded by this device I agree with him. They have to be excluded by a much more strict device. What is intriguing is whether we don't have here Laudan and Bloor colluding to endorse the old error of equating the rules of logic with the rules of thought (if any):

> Most philosophers are persuaded that reasoned and unreasoned beliefs represent two very different doxastic

> states and that these differences are due to different causal ancestries for the two types of belief (1981, p. 190) ... there are specific mechanisms for the generation of rational beliefs (among other [sic], various inferential mechanisms) which are not involved in the generation of unreasoned beliefs (1981, p. 187).

The apparatus of 'doxastic states' being caused by 'inferential mechanisms' seems a straight muddle of psychology and logic. We seem to be back at the idea that there is such a thing as straight and crooked thinking. How could we demarcate straight from crooked? Not by error!

> I think Bloor is quite right in asserting, with respect to our theoretical beliefs, that their truth status is largely if not entirely irrelevant to their explanation (1981, p. 186).

This is a relief, since most philosophers are probably persuaded that even if we possess the truth we do not know it, so to use it as a criterion to demarcate straight from crooked thinking would be futile. Bloor would probably go for a success-criterion based on prediction (see below). But in the absence of truth one man's judgement of predictive success may be another's judgement of failure -- so only consensus of the hard-to-define and circularly legitimated scientific elite will do.

Laudan's view seems to be that rational beliefs are those we have good reasons for. This is flimsy ground on which to rest claims about the rationality of the sub-culture known as science, still more the claimed rationality of a whole society compared to another (Laudan 1981, pp. 190-1) when it would involve the further mistake of treating social phenomena as reducible to the sum of the psychology of individual persons:

> Those in the rational society will adopt beliefs only after they have been carefully scrutinized and only after the agents have engaged in a specifically self-conscious causal interaction with relevant bits of the world (1981, p. 191).

Laudan's empiricist/inductivist psychologism raises many

problems, not least of which is how we ever get to know that our 'causal interaction with relevant bits of the world' has misled us. But I think I have said enough to indicate why my view is that the perfectly routine functioning of our minds may lead to many different outcomes, most of them erroneous. Errors are only uninterestingly products of our mistakes, including mistakes of thought: the interesting ones, from which we learn, are those for which there seem to be good reasons, both at the time and, when historical research is thorough, now.

In place of Laudan's view why not put the idea that rationality is a norm but not a norm of thought, rather a way of organising institutions and practices? Bloor says that 'the most important aspects of science concern the making and testing of predictions' (1981, p. 201). For the sake of the argument let us assume that this is so. Science will be rational to the extent that it fashions institutions, adopts rules and practices that maximise its ability to achieve that goal of making and testing predictions. One could even add that the best making and testing of predictions will clearly stem from science that is true. But, by this criterion of Bloor's, astrology, which certainly aims to make and test predictions, will also be a science, possibly a more successful one than the physical sciences, since it favours formulation of its predictions in such a manner that the outcome of the tests is a foregone conclusion.

Popper, unlike Lakatos, says very little about rationality in connection with science; his principal discussions of it are in connection with the explanation of social action. In Popper's philosophy of the social there is a rationality principle which urges us to explain actions on the assumption that the actors are aiming at goals but that any difficulty we have in seeing this will be to do with our grasp of the situation of the actors, as it is and as they see it. He has pointed out again and again that Freud explains even neuroses this way. Hence I do not think Popper would go along with Laudan's idea of rational and irrational societies, with Lakatos' idea that intellectual debate is rational and social psychology is irrational, or any of this. Indeed, he was delighted with my attempts when studying anthropology to render the most bizarre kinds of conduct rationally intelligible (both that of primitive people and of students of primitive peoples) by application of the rationality principle and analyses of the situation. Popper's paradigm of rationality is, he says, critical disucssion and science is rational to him to the extent that it approximates a well-conducted critical discussion or, more precisely, to the degree to which its discussions are criticial, i.e. utilise rules to maximise intellectual risk and the search for refutations. Is there not some problem here, does not this amount to Popper's way of passing judgement on the rational/irrational issue? No. As I have explained already, Popper virtually empties the notion of irrationality of any content and abandons it because it can be used to explain everything and hence nothing. Secondly, while Popper holds up critical discussion as an embodiment of rationality he does not present us with a fixed end point of perfectly rational critical discussion to which we approximate or from which we fall short. Rather does he in *The Logic of Scientific Discovery* look at the most intellectually successful of all critical discussion traditions, that of science, and show how it can engage in a good deal of self-improvement (his suggested new methodological rules).[4] And it follows from this that Popper's notion of rationality is a regulative idea. Part of the effort and results of science consists of research into and learning about the practical embodiment of rationality in fashioning institutions, devising rules and practices which maximise boldness of ideas and of criticism. Science has an exalted status by this measure because among the many goal-directed and hence rational activities to be found in society it focusses on the goal of increasing our knowledge. Action based on knowledge is, other things being equal, more likely to achieve given goals than action based upon ignorance. Hence the general search for knowledge is a way of generally improving all our efforts to achieve our goals.

3. Mistakes About Science

Last but more by way of a bang than a whimper, Bloor and Laudan conduct their dispute almost entirely in terms of something they call 'belief'. They take it that explaining science consists of explaining scientific beliefs. The dispute over the limits of the sociology of knowledge is read as whether all or only some 'scientific beliefs' should be subject to sociological or other external explanation. It is because of his fixed gaze on beliefs that Laudan is forced to argue that when the clockwork is running properly only beliefs backed by good reasons are held. One can sympathise with Bloor's example of the psychologist Koffka who tries to explain perceptions and illusions by the same mechanism (1981, p. 205).

The subjectivism of our two contestants is naked and unashamed: science is what scientists believe (Barnes and Bloor 1980, n.5). Gone is the old Platonic idea that science is knowledge, episteme, and never should be confused with opinion, *doxa*. For a moment let us take this *prima facie* plausible yet fundamentally preposterous identification seriously. Beliefs are presumably states of mind, something like 'holding', in which what is held is ideas, that is, theories, facts and what not. Let us now consider scientist X who has (in his head?) many 'beliefs'. For example, if he is Sir Isaac Newton he has the beliefs of Plymouth Brethren; if he is Galileo Galilei he has the beliefs of a good Catholic of the time; if he is Sir James Jeans he certainly subscribes to a diffuse Christianity. I imagine that neither Laudan nor Bloor would wish to classify these religious beliefs as science. Yet they are the beliefs of scientists. How, then, are we to demarcate the beliefs of scientists that are scientific from those that are not? Could we demand that the beliefs be empirical, or falsifiable, or backed by good reasons?[5] Yes, but this would not do the trick. A belief, a state of mind, cannot be empirical, falsifiable or based on good reasons. Only the *idea* that is the content of the belief state, or what the believer 'grasps' or 'holds' can be empirical, falsifiable, or backed by good reasons. Both these criteria and the 'contents' of beliefs do not belong to scientists or their minds. In fact these ideas and these criteria for assessing ideas as scientific transcend the beliefs and practices of scientists altogether. Science defines scientists rather than vice versa. To take the scientific elite as a measure of science is circular.

This thought suggests another line of argument. When Einstein dreamed up his special theory he was a patents clerk in Switzerland, not a scientist. Sir Isaac Newton was the keeper of the Royal Mint. These and other examples could be used to create problems from the other direction. How can it be that manifestly scientific ideas can be believed, even invented and then believed, by non-scientists? Again, we have to turn to content and criteria for demarcating content for an answer.

Popper makes yet another argument that has force. 'Einstein proposed his theory of general relativity; ... he defended it patiently against violent criticism; ... he suggested that it was an important advance, and that it should be accepted as an improvement on Newton's theory; yet ... he never accepted it himself' (1963, p. 969). What this suggests is that there is a further subtlety. A scientist may declare that he doesn't believe in belief. That, if he has beliefs, these have no connection with his contributions any more than Hume could believe in his refutation of induction, possibly one of the greatest contributions to philosophy anyone ever made.

With this I will drop the pretence of taking seriously identification of science with scientists' or anybody else's beliefs. It is not clear what they are, it is not clear how to demarcate them, and, logically, they are irrelevant to science. The temptation is at this point to try once more to explain the value, for all its problems, of Popper's notion of World 3. But that tends to be futile and off-putting, I find. Instead let me suggest the objectivist picture of science by different means. First point: science is a system of statements, not of beliefs, practices or elites: those latter are merely parasitic on the former. But

> Science is not a system of certain, or well-established statements; nor is it a system which steadily advances towards a state of finality. Our science is not knowledge (episteme): it can never claim to have attained truth, or even a substitute for it, such as probability.... Like Bacon, we might describe our contemporary science -- 'the method of reasoning which men now ordinarily apply to nature' -- as consisting of 'anticipations, rash and premature' and of 'prejudices' ... Our method of research is not to defend them, in order to prove how right we were. On the contrary, we try to overthrow them ... in order to put forward, in their stead, new unjustified and unjustifiable anticipations, new 'rash and premature

> prejudices', as Bacon derisively called them ... Bold ideas, unjustified anticipations, and speculative thought, are our only means for interpreting nature: our only organon, our only instrument, for grasping her (1959, pp. 278-80).

Now it should be apparent why I claimed that both Bloor and Laudan agreed on matters with which one could take issue. Each has been led off by red herrings, whether the logical muddle that psychological and other considerations can throw light on the logical relations between the systems of statements that constitute science, or the idea that there is some marvellous clockwork way of thinking that leads us to the right conclusions. By Bloor's lights Popper is ignoring all the intricate connections of science as just another social institution to its peers; by Laudan's lights Popper is endorsing as part of science precisely the kind of thought he would call malfunctioning and consign to the realm of the irrational.

At the beginning of this paper I commented on their joint agreement to dismiss the transcendental and I refrained from criticising them too severely for taking Imre Lakatos' epicyclical travesties of fallibilism as a guide to Popper's philosophy of science. Now, however, the gloves have to come off. Tired old philosophies of inductivism, which identify science with a kind of right thinking, right opinions and right beliefs really cannot be coherently recycled in this day and age. But most important of all, neither transcendental considerations nor metaphysics can be avoided in the way that both our opponents would love. The metaphysics of positivism are well known, as are the weaknesses. Bloor's innocent description of himself as an inductivist suggests he knows and doesn't care. He takes the trouble to argue with Laudan and others, but when backed into a tight corner inconsistently complains of his opponent's 'verbal bias'. That sounds to me very much like the argument 'I know there's an answer to you but for the moment I just can't seem to find the words....' Laudan is a more subtle case. It took me thirteen closely argued pages to show that his book is riddled with unconscious inductivist and positivist assumptions. What shows in the present dispute is his espousal of belief-philosophy and the consequent morass he gets into, especially the muddle about right thinking. So,

> "A plague o' both your houses." (Shakespeare, *Romeo and Juliet*, III, I, 112.)

Department of Philosophy
York University

Notes

1. "Mannheim's understanding of the sociology of knowledge was much more far-reaching than Scheler's ... Society was seen as determining not only the appearance but also the content of human ideation, with the exception of mathematics and at least parts of the nature sciences." (Berger and Luckmann 1966, p. 8).
2. In the penultimate footnote to their (1980), Barnes and Bloor misread an attempt of mine (1979, pp. 496-7) to give a sense of awe to the scientific achievement and to suggest this is a down-to-earth explanation of the yearning for contact with the divine, as an expression of religious faith! They had no reason to know I am an atheist, but the text was nearer to sacrilege than to religious expression.
3. I refer to chapter 1, 'Popperians versus Hegelians', of his forthcoming Rationality and Relativism in the Social Sciences.
4. Mark Blaug makes a useful attempt to extract these from the text (1980, ppp. 18-19).
5. Laudan: "no one, philosopher or sociologist, has yet set out any acceptable account of what cognitive or methodological features demarcate the sciences from the non-sciences." (1981, pp. 180-181) The amount packed into the word 'acceptable' here is quite unacceptable.

Bibliography

Barnes, Barry and David Bloor: 1980, 'Relativism, Rationalism and the Sociology of Knowledge', MSS.

Berger, Peter and Thomas Luckmann: 1966, *The Social Construction of Reality,* Doubleday, New York.

Bloor, David: 1981, 'The Strengths of the Strong Programme', *Philosophy of the Social Science,* 11, 199-213.

Blaug, Mark: 1980, *The Methodology of Economics,* Cambridge University Press, Cambridge.

Gellner, Ernest: 1964, *Thought and Change,* Weidenfeld, London.

Gellner, Ernest: 1975, *Legitimation of Belief,* Cambridge University Press, Cambridge.

Gellner, Ernest: 1982, 'The Paradox in Paradigms', *Times Literary Supplement,* April 23, no. 4125, 451-2.

Jarvie, I.C.: 1972, *Concepts and Society,* Routledge, London.

Jarvie, I.C.: 1979, 'Laudan's Problematic Progress and the Social Sciences', *Philosophy of the Social Sciences,* 9, 484-97.

Jarvie, I.C.: 1981, 'The Rationality of Creativity', in Denis Dutton and Michael Krausz (eds.), *The Concept of Creativity in Science and Art,* Nijhoff, The Hague, pp. 109-28.

Kuhn, T.S.: 1962, *The Structure of Scientific Revolutions,* University of Chicago, Chicago.

Kuhn, T.S.: 1970, 'Logic of Discovery or Psychology of Research?', in I. Lakatos and A. Musgrave (eds.), *Criticism and the Growth of Knowledge,* Cambridge University Press, Cambridge, pp. 1-23.

Lakatos, Imre: 1976, *Proofs and Refutations,* Cambridge University Press, Cambridge.

Laudan, Larry: 1977, *Progress and Its Problems,* University of California Press, Berkeley and Los Angeles.

Laudan, Larry: 1981, 'The Pseudo-Science of Science?', *Philosophy of the Social Sciences,* 11, 173-98.

Laudan, Larry: 1982, 'More on Bloor', *Philosophy of the Social Sciences,* 12, 71-4.

Popper, K.R.: 1959, *The Logic of Scientific Discovery,* Hutchinson, London.

Popper, K.R.: 1963, 'Science: Problems, Aims, Responsibilities', Federation Proc., Federation of American Socieities for Experimental Biology, 22, Part I, 961-72.

Popper, K.R.: 1970, 'Normal Science and Its Dangers', in I. Lakatos and A.M. Musgrave (eds.), *Criticism and the Growth of Knowledge*, Cambridge University Press, Cambridge, pp. 51-8.

Popper, K.R.: 1972, *Objective Knowledge*, Oxford University Press, Oxford.

Schilpp, P.A. (ed.): 1974, *The Philosophy of Karl Popper*, Open Court, La Salle (Ill.)

Andrew Lugg

TWO HISTORIOGRAPHICAL STRATEGIES: IDEAS AND SOCIAL CONDITIONS IN THE HISTORY OF SCIENCE

Science has generally been thought to be an intellectual enterprise without equal and its history to be best explained using the techniques of the intellectual historian. For most historians of science, the work of Alexandre Koyre, I.B. Cohen, Henry Guerlac and Marshall Clagett, which focuses on scientific ideas and ignores what happens in society at large, provides the model of good historical practice.[1] However, a significant number of historians now insist that studies which leave the social context unexamined are incomplete and unilluminating. For such historians, the history of science should be largely a matter of relating scientific ideas to the social context in which they occur. Why is this? And what reasons do these "social historians" have for rejecting the older historiographical tradition?

I

Critics of orthodox history of science support their views by pointing to a growing body of case studies informed by a sociological approach to scientific knowledge. They set forth their challenge to intellectual history by arguing that the case studies now available establish what Steven Shapin has called "an historical sociology of scientific knowledge".[2] The main claim is that once we put aside "superficial and unconstructive characterizations of what sociological explanation is",[3] we shall see that intellectual history of science must be supplemented if not replaced. However, when we examine the social historians' case studies in conjunction with the various claims that are made on behalf of these studies, social history no longer appears to be antithetical to intellectual history.

J. R. Brown (ed.), Scientific Rationality: The Sociological Turn, 183–208.

What kinds of study are said to establish the importance of examining science in sociological terms? The following examples serve to focus the discussion:

(1) Theodore Brown has argued that in the second half of the seventeenth century members of the Royal College of Physicians adopted iatromechanism to turn back a threat to the College. In Brown's view, the collegiate physicians accepted iatromechanism not only "when the College had socio-political troubles" but also "because (it) had them".[4]

(2) Similarly, Paul Forman has argued that in the 1920s German physicists embraced the anti-mechanistic and anti-deterministic ways of thinking then prevalent in Germany to further the interests of the physics profession in what was at the time a hostile environment.[5]

(3) According to Martin Rudwick, Poulett Scrope's work in politics and economics played a significant role in his adoption of geological actualism. For Scrope, an early nineteenth-century British geologist, geological time was analogous to money in the banking system, "there (being) no intrinsic limit to the supply of time, any more than there is to the supply of money".[6]

(4) Robert Young has detailed the influence on Darwin of Malthus's ideas about the survival of societies and the ideas of commercial breeders about selecting for fitness.[7] And Michael Mulkay has suggested on the basis of Young's work that Darwin's appeal to the analogy between natural and directed breeding reveals that he had a "strong commitment to the perspective of the breeders".[8]

(5) Steven Shapin has traced the dispute between phrenologists and antiphrenologists in early nineteenth-century Edinburgh to the differing social positions of the members of the two groups. As Shapin presents the case, the debate arose primarily because the phrenologists were outsiders whereas the antiphrenologists were members of the Edinburgh establishment.[9]

(6) Finally, Donald MacKenzie has suggested that Karl Pearson's adoption of continuity theories of biological and social change reflected "with exceptional clarity the social interests of the professional middle class to which he belonged".[10]

These studies differ from each other in various respects. In some cases the emphasis is on the kinds of considerations that scientists invoke. Thus, Young stresses that Darwin adopted his views in part because he subscribed to Malthus' social conceptions and because he brought to bear analogies with phenomena in other areas. In other cases, scientists are seen as adopting theories to further social goals. For instance, Brown takes himself to have shown that the collegiate physicians adopted a new view in answer to a challenge to their professional prestige. And in yet other cases, social position and social interests are held to have influenced the types of views scientists adopt. Thus, Shapin argues that an explanation of the debate between phrenologists and antiphrenologists should "not stop with the identification of methodological and conceptual incommensurables (but should go on to) find the source of (the) conflict and trace its roots to social conflict, differing social experiences, values and interests".[11] Furthermore, notice that it is the general form of theories, not their details, that is usually held to have been prompted by social considerations. Few social historians now insist that every aspect of a theory can be correlated with some social role or structure. Thus, MacKenzie's claim that Pearson's preference for continuity theories reflected his social interests should not be taken to imply that the specific continuity theory that Pearson developed was also prompted by other than purely scientific considerations.[12]

Common to all six explanations, however, is an implicit rejection of the kind of social history often attributed to an earlier generation of social historians. Although the explanations differ, none is plausbily interpreted as asserting that scientific ideas directly reflect underlying social realities. Social factors are seen as playing a subsidary, albeit a crucial role; they are thought of as "mediating" rather than as determining ideas.[13] Unlike some of their predecessors, modern social historians are sensitive to the argument that ideas are too numerous, subtle and fleeting to be mere reflections of the relatively coarse differentiations characteristic of social structures. Social history in its latest form thus cannot be criticized on the grounds that similar scientific developments have occurred in different social environments.[14]

This is a particularly important point since it is only by stressing the indirect role of social factors that social historians can avoid the charge that their sociological analysis impugns the "rationality of science". The problem with social history as it sometimes used to be done was that it pictured scientists as being engaged in wholesale rationalization; it had the effect of writing the mind out of history and of making social structures seem capable of conjuring up scientific ideas on their own. By contrast, modern social historians see their kind of historiography as being perfectly compatible with the conception of science as a paradigm of rationality. As Barry Barnes puts the point in a recent book, "the reader must simply understand that in asserting the insufficiency of 'reason' in science, the text will in no way imply that scientists are unreasonable men".[15]

If we take such disclaimers seriously -- and it seems to me that we must -- we evidently cannot interpret the social factors in the cases mentioned as crude social pressures or blind social interests. For that would mean that rationalization in science is the rule rather than the very rare exception. In particular, if we agree that the social historian's key case studies involve no unreasonable procedures, the collegiate physicians cannot be held to have adopted iatromechanism in order to garner support from society at large; Darwin cannot be seen as having attempted to further the interests of, say, the commercial breeders; and the Edinburgh phrenologists and antiphrenologists cannot be thought of as having developed their views with their social positions in mind. One cannot hold both that scientists were "reasonable men" and that they were rationalizing underlying social interests.

How then should we interpret the social historian's case histories? What exactly does the evidence, as opposed to its interpretation, actually establish? Consider first Scrope's analogy between time and money. The evidence provided certainly does not establish that Scrope was furthering anyone's interests; it rather leaves open the possibility that Scrope believed his analogy to be capable of serving as a premiss in an argument for geological actualism. Perhaps Scrope had ulterior motives, but what is shown is only that he made use of his knowledge of economics. And similarly for Darwin: the evidence offered is compatible with the assumption that he took Malthus's observations about societies and the analogy between "natural" and "directed" breeding to be insights that supported the truth of evolutionary theory. In both cases, social factors can be seen as having played no more than a trivial role. We are given no reason to forego the unpretentious view that "social" conceptions were invoked simply because they were -- quite reasonably -- held to have a bearing on the truth of the theories under investigation.[16]

As for case histories such as Brown's and Forman's, saying that scientists were prompted to reconsider and even revise their views because of outside pressure does not mean that they adopted these views for social reasons. It is entirely consistent with Brown's evidence that the collegiate physicians accepted iatromechanism because they thought it provided a way of reconciling older Galenic medicine with the new findings of William Harvey. As Brown himself concedes, "advocating or accepting iatromechanical ideas could well have seemed to many or most of those involved like the rational rather than the expedient thing to do, especially since iatromechanism also satisfied a real intellectual need of the collegiate physicians".[17] And in the case of the German physicists in the 1920s, the evidence that Forman develops can equally well be seen as suggesting that the physicists' recognition of the anti-mechanism and anti-determinism then being advocated resulted in their appreciating more fully the possibility of coherently introducing indeterminism into physics. if this is right, the problem of how we can insist that these scientists proceeded reasonably immediately vanishes.

Finally, we need note that the evidence presented in Shapin's and MacKenzie's studies is insufficient to establish that scientists from different social groups adopted different theories because they were guided in their deliberations by different social interests. All that Shapin and MacKenzie in fact demonstrate is that their subjects' deliberations had social preconditions. The evidence that Shapin provides only shows that the phrenologists and the antiphrenologists had different background knowledge as a result of their occupying different social positions. They disagreed not because they had different social interests but because they took different bodies of information to be true. Social factors may have occasioned their having different bodies of background knowledge, but this is of no significance: scientific knowledge is always occasioned by social factors. And likewise in the case of Pearson's commitment to biological and social gradualism. MacKenzie admits that the interests he mentions are "theoretical posits", and the evidence he provides can be seen to show that Pearson's views were shaped for the most part by the various intellectual influences to which he was subject as a result of his being a member of the professional classes.[18] In both cases, then, the social factors cited are best seen as playing a secondary and rather trivial role.

These observations provide a way of reconciling the claim that scientists are reasonable with the claim that their appraisals are "influenced" by social factors. But they also undercut the alleged novelty of the social historian's approach. For intellectual historians have long emphasized the crucial role of metaphors and other "extraneous" considerations in the development of science; they have long argued that there are no particular kinds of consideration that prompt scientists to formulate and appraise theories; and they have long taken it to be self-evident that scientific knowledge has social pre-conditions. One need only recall that intellectual historians themselves have stressed the importance in science of metaphors such as Descartes' comparison of the way light travels to the way a blind man finds his way around with a cane; that they have argued at length that social concerns prompted much scientific work since the seventeenth century; and that they have never denied that the very existence of science depends on social preconditions such as "a certain stability in some part of society, a certain degree of literacy, and a certain desire for intellectual pursuits".[19] Thus, if we view the social historians' accounts as suggested, social history becomes to all intents and purposes a branch of intellectual history.

To sum up the argument so far: To view social history as challenging the "rationality of science" runs counter to the claims that social historians make on behalf of their studies.[20] But if we take social historians to be holding that their subjects proceeded in a completely proper way, we must view metaphors as "scientific factors" and social circumstances as occasioning rather than determining ideas. Thus, in sofaras the cases I have been considering are typical -- and it seems to me that they are -- we must reject Shapin's claim that social history can be established by examining case histories. Even when we put aside "superficial and unconstructive" conceptions about what is and what is not social, social history does not appear to be superior to intellectual history as it has been traditionally pursued.

II

If the case studies that social historians provide are social only in an entirely uncontroversial sense, why are they held to pose so serious a challenge to intellectual history? Why do social historians insist on the need for a fully sociological account of scientific knowledge even when the historical material they provide can be readily reinterpreted within the framework of the intellectual historian? An examination of the literature reveals two main reasons for this. First, social historians take themselves to have good reasons for thinking that sociological analysis is appropriate for every scientific belief; second, they believe that they can show that intellectual history is unscientific.

A. Social Historians on Social History.

One reason why many social historians think that social history is indispensable relates to the distinction between factors that are "internal" and factors that are "external" to science. Noting that even the best scientists often appeal to "external" as well as "internal" considerations, some social historians have concluded that scientific developments must be explained in sociological terms. Consider Scrope's use of banking metaphors or Darwin's use of Malthus's ideas and the metaphor of natural selection. Since these considerations were clearly external to geology and biology, their cogency -- according to this line of thought -- cannot be "rationally" explained. Rather, we should dispense with the distinction between internal and external factors and recognize that all reasons should be explained sociologically.[21]

Yet recognizing that scientists legitimately appeal to external considerations lends no support to the social historian's position.

For it has long been one of the main themes of intellectual history that considerations alleged to be external play a major role in the development of science. In pointing out that scientists appeal to metaphors, analogies and philosophical, political and social conceptions, intellectual historians have not seen themselves as going beyond the bounds of intellectual history properly understood. They acknowledge that nonscientific ideas are ideas nonetheless. The mistake of the social historian is to see all external factors as requiring social analysis. On the one hand, they reject the traditional distinction between internal and external factors; on the other, they tacitly go along with the tradition and hold that only an exceptionally narrow range of considerations can be treated without invoking social factors of one sort or another.

Besides seeing the demise of the internal-external distinction as supporting their view, proponents of social history believe it to be vindicated by recent developments in the philosophy of science. Thus, David Bloor takes the fact that theories are always underdetermined by the data on which they are based to establish the conclusion that "the theoretical component of knowledge is a social component".[22] Barry Barnes invokes the thesis that "nothing in the nature of things, or the nature of language, or the nature of past usage, determines how we employ, or correctly employ our terms" in support of the claim that "all instances of concept application" require sociological analysis.[23] And Mary Hesse argues from the widely accepted view that there are no criteria of evalution applicable in all conceivable circumstances to the conclusion that scientific beliefs should always be analyzed in sociological terms.[24]

Such inferences are common, but dubious nevertheless. The various philosophical points may be correct, but they hardly establish the need for a social analysis of scientific knowledge. Social historians all too often fail to see that the philosopher's views can be accommodated at least as well within the intellectual historian's framework: we are not forced to choose between extremely implausible philosophical views and a full-blown sociology of knowledge. In particular, Bloor overlooks the fact that scientists routinely and quite properly bridge the gap between theory and data by invoking metaphors, analogies, philosophical ideas, methodological principles and other "intellectual" considerations.[25] Barnes mistakenly takes the arguments of Wittgenstein and others against traditional philosophical conceptions of meaning to show that there is no such thing as meaning or deliberation in the usual, non-philosophical sense.[26] And those like Hesse who argue for "relativism" and against "rationalism" fail to appreciate sufficiently that scientific inquiry can occur in the absense of fixed rules, and that intellectual history itself is "relativistic" in just the sense endorsed by social historians.[27]

Lastly, and most importantly, many social historians seem to have become committed to social history because of the ambiguity of "social" and similar words. Recognizing that science is naturally and quite properly discussed in social terms, they mistakenly conclude that scientific knowledge must be analyzed sociologically. Thus, David Bloor has observed that "science is a social phenomenon, so we should turn to the sociologist of knowledge", and Everett Mendelsohn has argued that "science is an activity of human beings acting and interacting, thus a social activity. Its knowledge, its statements, its techniques have been created by human beings and developed, nurtured and shared among groups of human beings. Scientific knowledge is therefore fundamentally social knowledge".[28] In both cases, the premisses are plausible only when "social" means "having to do with a number of individuals", yet the conclusions require that we understand "social" in the sense of "having to do with certain kinds of pressure, interest and the like". Put otherwise, arguments like Bloor's and Mendelsohn's are plausible only because equivocation on "social" is all too easy and all too inviting.

Less blatant but no less problematic is the claim that social analysis is required because "authority" and "patronage" play important roles in science and because scientists draw on "cultural resources", have "social interests", make "political moves" and "negotiate with one another". Such claims are true only if the concepts of "authority", "interests", "negotiation" and so on are used in a weak sense, but for the conclusion to follow they must be taken in a full-blooded sociological sense. It is true that scientists draw on cultural resources in the sense that they make reference to generally accepted views, that they are politic in the sense that they attempt to forestall criticism and to present their views in an attractive manner, and that they negotiate in the sense that they debate among themselves. But this does not in the least show that scientific knowledge must be analyzed in sociological terms. To estalish this, it must be shown that in drawing on cultural resources, scientists are attempting to further the interests of the gorup that has these resources; or that in making political moves, they are behaving like politicians; or that in negotiating among themselves, they are wheeling and dealing, not just debating.

In general, the problem is another example of what Dennis Wrong has referred to as "the oversocialized conception of man in modern sociology".[29] So much is seen as social that the notion becomes practically useless, and sociological analysis becomes so diffuse and so all-encompassing that virtually all history becomes social history. There can be no doubt that science is a social endeavor; but that is nothing any intellectual historian ought to dispute.

B. Social Historians on Intellectual History

The social historian's second point is that intellectual history is marred by unempirical assumptions. In the literature, it is frequently suggested that only social history is entirely empirical and that we have no option but to attempt to develop a fully social history of science.[30] In particular, it is alleged that intellectual history is inadequate because (a) it does not deal with concrete events, (b) it assumes non-causal interactions between ideas, (c) its methods are not those of the natural sciences, and (d) its explanations depend on judgements about the rationality and the value of the views under investigation. In other words, whereas social history is "materialistic", causal, "naturalistic" and "non-normative", intellectual history is said to be "idealistic", non-causal, non-naturalistic and normative.

This criticism is certainly invited by the way that intellectual history is presented by some of its proponents. Intellectual history would seem to assume abstract entities and non-causal relations if -- as is often argued -- scientific ideas follow their own "inner logic" and are "logically or methodologically (reducible) to some pattern".[31] Its method is not the method of the natural sciences if it consists of the "history of rational thought about nature ... requiring for its understanding only the attempt on the part of the historian to 'think the scientist's thought after him'".[32] And it cannot be entirely descriptive if "history of science without philosophy of science is blind" or the province of intellectual history is to explain "well-founded" beliefs.[33]

However, loose programmatic remarks cannot serve as a basis for rejecting intellectual history, let alone for establishing the social historian's own controversial approach. In defense of their view, intellectual historians can justifiably respond they are not treating scientists' ideas as disembodied, Platonic entities which can only be grasped or "intuitively apprehended". The ideas studied by Koyre, Cohen, Clagett and other intellectual historians are not in the least disembodied; they are ideas espoused by historical subjects at particular times and places. The intellectual historian who asks what Newton means when he speaks of universal gravitation is being no less "materialistic" than the social historian who attempts to link these ideas to Newton's general social milieu. Newton's formulating the law of universal gravitation was as much a concrete occurrence as his being born on Christmas day during the Protectorate. In fact, even the history of historians like Arthur Lovejoy and J.R. Partington, which approximates the image of disembodied history, can be readily retrieved by the simple expedient of taking every idea to be "fully embodied".

Intellectual historians, moreover, should not be seen as providing explanations in terms of reasons rather than causes, though some have mistakenly thought their concern to be with "non-causal rational explanation".[34] For what they are in fact attempting to show is how scientists come to certain beliefs as a result of their having accepted other beliefs as reasons. Whether or not reasons are causes, a scientist's accepting a reason can undoubtedly be a cause of his or her coming to hold a new belief.[35] When intellectual historians explain Newton's acceptance of the law of gravitation, they do not merely show that the law is logically related to other ideas that Newton held; they also show that these very ideas prompted him to accept the law. For the intellectual historian, then, what counts are the fully embodied beliefs of scientists and the fully causal reasoning processes that they actually employ.

But this still leaves open the question of whether intellectual historians have to invoke special methods, and in particular whether they must engage in a highly suspect form of mind-reading. In the interests of good scientific practice, should not historians avoid imputing internal states of belief, motive and intention to their subjects?[36] Certainly, it must be granted that ascriptions of beliefs to scientists, no less than to anyone else, are in some sense speculative. But there is surely no reason why they should be any less under empirical control than speculations in the behavioural and social sciences. Social historians too must interpret historical documents to determine what their subjects were thinking, and past social causes are not any easier to establish than past beliefs and intentions. To insist that such ascriptions should be made with care is one thing, to derogate them as mind reading quite another.

Nor, finally, should it be thought that intellectual history presupposes philosophical or normative considerations in a way that social history does not. True, intellectual historians mention scientists' reasons, but this by itself does not establish that they make normative judgements. Intellectual historians do not typically attempt to depict science as an unswerving march towards the truth; they need not be apologists for science, nor mere chroniclers of its lasting achievements. When they cite a reason, they need not praise or blame it, nor need they imply that it conforms to a particular philosophical theory of rationality. An intellectual historian's account of how Newton justified his acceptance of universal gravitation may have little to do with how we would go about justifying it or with how Newton himself ought to have justified it. For the intellectual historian, the crucial point is that Newton embraced the view that he did because he accepted certain considerations as reasons, because he -- Newton -- thought these to be sufficient to establish his conclusion. In general, when intellectual historians speak of reasons, they speak of them in a purely descriptive way. They study the "rationality" of beliefs only insofar as this concerns what their subjects take to be reasonable.

The intellectual historian's approach is thus far more sophisticated than many social historians allow. Intellectual history, no less than social history, can be empirical in the strictest sense: the charge that intellectual historians are unempirical idealists applies to a small group of historians at most. In fact, it is rather the social historians themselves who proceed unscientifically. For it is they who stipulate in advance that scientific beliefs should be explained in a particular manner. It is they, not intellectual historians, who make judgements prior to empirical investigation concerning the type of explanation required. Good scientific practice, by contrast, leaves open the question of whether a belief should be explained by citing reasons, social causes or some third possibility.[37]

III

Intellectual historians, whose views I have been defending, hold that scientific beliefs should be socially explained only when scientists accept them as a result of social pressure or in order to further their social interests. Although all beliefs have social preconditions, explanations in terms of reasons are seen to pre-empt the need for social accounts. When scientists accept beliefs on the basis of reasons, it is unnecessary and inappropriate for the historian to go beyond these reasons and examine the social contexts in which they occurred. Yet in many cases social explanations appear to shed light on scientists' acceptance of beliefs even when these were accepted for reasons. In particular, social analysis can enhance our understanding of Scrope's use of banking metaphors, Darwin's appeal to Malthusian conceptions, and the debate between phrenologists and antiphrenologists. The intellectual historian may be right to point out that these scientists were not influenced by social pressures or interests, but it may nevertheless be helpful to relate their ideas to their social circumstances. As the social historian insists, it is occasionally appropriate to stress not only scientists' ideas but also the social circumstances in which these ideas are formulated and defended.

The point is not that explanations are always "incomplete" in that further explanation can always be provided. Social historians who emphasize the social contexts of scientific beliefs are not simply elaborating perfectly adequate explanations; they are not just pointing out that certain beliefs have certain social preconditions. Rather we should see them as claiming that on occasion no explanation will be adequate unless it goes beyond the beliefs of the scientists involved to the social circumstances surrounding those beliefs. On this view, explanations of the beliefs of Scrope, Darwin and the phrenologists in terms of reasons are incomplete not because they can be supplemented, but because they are insufficient, inadequate, unrevealing.

When explaining historical episodes, our aim is to account for them in our own terms, not those of the scientists involved. We embark on our historical deliberations armed with expectations and information that differ considerably from the expectations and information of our subjects. Taking seriously the pragmatics of explanation,[38] we cannot assume that the answer to a particular question about a scientific belief must be of a particular kind. The answer will rather depend on what we know and what we find puzzling. Given one body of background information, an explanation emphasizing certain features of the event may be appropriate; given other information, quite different features might have to be stressed. If we take explanation to be an "objective" relation between one statement (the explanadum) and others (the explanans), no such variability will arise. But as soon as we see explanation as a process of enhancing understanding, we can no longer set aside the possibility that we must go beyond the scientist's beliefs however reasonably these were acquired. The way we proceed must depend on what makes sense to us as well as on what our historical subjects actually did.

These remarks do not vindicate the strong claim that social history is always required. In many cases we can explain to ourselves why scientists adopted particular theories simply by noting the reasons they gave. Thus, Newton's adotpion of universal gravity, Lavoisier's formulation of the oxygen theory of combustion and Maxwell's acceptance of electromagnetic theory can in all likelihood be explained without considering their social circumstances. In such cases we can explain to our own satisfaction how scientists proceeded by adopting the standpoint of the scientists themselves. On the other hand, it would be a mistake to jump to the conclusion that a social explanation is required wherever it is appropriate to adopt the point of view of an outside observer. For we can sometimes explain why scientists adopted certain views by relating these views to nonsocial circumstances. We may, for instance, be able to account for a certain scientist's position simply by noting that he or she was misled by a deceptive phenomenon. Nevertheless, when explanations go beyond a scientist's beliefs, they often go beyond them to his or her social circumstances.

The possibility of explaining reasonable scientific beliefs in social terms can be illustrated by considering Scrope's use of banking metaphors and Darwin's appeal to Malthus's ideas about society. We can reconstruct Scrope's and Darwin's reasoning in detail, outlining both the evidence and the methodology to which they appealed. Yet the question remains of why they took the reasons they gave to be reasons. For us, Scrope and Darwin may appear to be reasoning in controversial, odd or questionable ways: what appeared reasonable to them need not appear reasonable to us. However, if we relate their ideas to their social circumstances, we may be able to make better sense of their inquiries. Suppose that it was only in the early nineteenth century that the information about banking invoked by Scrope became "cultural resources". We could then clarify Scrope's use of the analogy between geological time and money by noting how developments in society had made available a new type of consideration. Likewise, suppose that Darwin had been impressed (as Wallace seems to have been[39]) by the toll that wars, famine, sickness and the like take of human life. We could then explain Darwin's adoption of Malthus's ideas by noting that the social world with which Darwin was familiar provided him with a model of the biological world. In other words, the explantion of why Scrope and Darwin adopted certain considerations as reasons could be in part that they lived in a certain type of society.[40]

Alternatively, consider the debate between phrenologists and antiphrenologists in nineteenth-century Edinburgh. To explain this debate we cannot simply explain why the phrenologists accepted their views and why the antiphrenologists accepted theirs. What needs to be accounted for is why phrenologists accepted one theory <u>while</u> the antiphrenologists accepted another. We require an explanation of a contrast, not separate explanations of the two contrasted items.[41] This recognized, it becomes clear that an account tracing the reasons of the two sides to their social positions would be explanatory in a way that accounts of the two sides' reasoning would not. We may agree that social factors only occasion the views of the phrenologists and antiphrenologists (see section I), yet still insist that an adequate explanation of their disagreement must make reference to the social positions that occasion their differing bodies of belief. One might, for instance, argue that the social positions of the phrenologists and antiphrenologists made available or delimited their background information in different ways; or one might argue that the phrenologists based their arguments on their experiences as outsiders while the antiphrenologists based their arguments on their quite different experiences as members of the Edinburgh establishment.[42]

In such explanations -- and no doubt other kinds of case are also possible -- scientific beliefs are related to social circumstances, even though they are not rationalizations. We appeal to the social contexts in which ideas are appraised, but we do not in any way imply that social interests or social pressures influenced these appraisals. Social factors are held to function as social circumstances, not as social causes. To employ a phrase of Cheyney Ryan's, the social factors are beyond the beliefs, not behind them.[43] The point is that we can better understand why Scrope, Darwin and the phrenologists reasoned as they did if we set their deliberations in wider social contexts. We do not question the claim that they adopted their views because they believed them to be true; nor do we challenge the contention that we can rethink their thoughts in the way advocated by intellectual historians.

It might be objected to this that we inevitably impugn the "rationality" of beliefs if we go beyond them to the surrounding social circumstances. Scientists should not be influenced in their acceptance and rejection of theories by their social environments. If they always proceeded in an appropriately scientific manner, social factors would never affect their decisions. In particular, Scrope should not have relied on generally accepted belief; Darwin should not have generalized from a few observations about human society to the nature of all biological groupings; and the phrenologists and antiphrenologists should not have confined themselves to the limited information they had as a result of occupying particular social positions. Indeed, given the historical accounts I have been asusming, is it not more plausible to see Scrope as arguing from authority, Darwin as hastily generalizing, and the phrenologists and antiphrenologists as committing the fallacy of provincialism? Why think that what has been shown is that Scrope and the rest were proceeding reasonably rather than rationalizing their social concerns?

To be sure, Scrope, Darwin and the others may have been responding to social pressure or attempting to further their social interests. Only historical investigation can decide that. However, the main point of the objection -- that social explanations of beliefs inevitably impugn their "rationality" -- must be rejected. For it is entirely compatible with the view I have been sketching that Scrope, Darwin and the phrenologists and the antiphrenologists were all acting in a perfectly proper manner. One can argue from generally accepted belief without "arguing from authority"; one can generalize on the basis of limited experience yet not "hastily generalize", and one can rely on information to hand without being "provincial". If Scrope and the others investigated the issues that concerned them in accordance with the generally accepted standards of the time, we can hardly criticize them for not having good reasons for accepting their views. Scientists cannot be expected to be omniscient. Their proceeding reasonably does not preclude the possibility that the availability of their information depends on developments in society at large; that facts about society have an important bearing on how they should consider non-social phenomena; or that their social positions circumscribe the information to which they have access.

Lastly, it is perhaps worth comparing the conception of social history that I have been urging with a conception that Shapin sees as dominating the thinking of many critics of the sociology of knowledge. This conception, which Shapin calls "the coercive model", comprises the following six theses: (i) "sociological explantion consists in claims of the sort: 'all (or most) individuals in a specified social situation will believe in a specified intellectual position'", (ii) "the social (can be derived) by aggregating individuals", (iii) "the connection between social situation and belief (is) one of 'determination'", (iv) "the social (is the same as the) 'irrational'", (v) "sociological explanation (involves the) invocation of 'external' social factors", and (vi) "sociological explanation (runs counter to) the contention that scientific knowledge is empirically grounded in sensory input from natural reality".[44] By contrast, according to the view presented in this paper, (i) beliefs are unlikely to be related in a lawlike fashion to social situations, (ii) contrasts are not normally explained by conjoining explanations of the contrasted items, (iii) determinism may or may not hold, (iv) rational belief is sometimes appropriately explained in social terms, (v) sociological explanatioin may involve social considerations internal to science, and (vi) empiricism in some reasonable form may well be true.[45]

In my view, then, there is merit to both the intellectual and the social historian's positions, yet neither position is wholly adequate. Social historians are right to insist that reasonably held beliefs may be explained in sociological terms; they only err in assuming that social explanation is always required when in fact it is only infrequently appropriate.[46] On the other hand, intellectual historians are justified in emphasizing both the preeminent role of reasoning in science and the fact that social factors generally do no more than occasion and circumscribe beliefs; they can only be faulted for restricting the techniques of social explanation to cases in which beliefs are adopted to further social interests or as a result of social pressures. Recognizing that explanations of scientific beliefs, like explanations of anything else, are always relative to our concerns and interests, we can transcend both historiographical strategies. We can retain the social historian's main idea even as we preserve the core of the intellectual historian's approach.[47]

Department of Philosophy
University of Ottawa

Notes

1. See, for example, Koyre (1965), Cohen (1980), Guerlac (1975) and Clagett (1959).
2. Shapin (1982), p. 158.
3. *ibid.* For more on what Shapin regards as a "superficial and unconstructive" characterization of social history, see the remarks about the "coercive model" at the end of this paper.
4. Brown (1970), p. 29.
5. Forman (1971).
6. Rudwick (1974), pp. 240-241.
7. Young (1969) and (1971).
8. Mulkay (1979), p. 104.
9. Shapin (1975) and (1979).
10. MacKenzie (1979), p. 125. To avoid misunderstanding, I should note here that many other types of study have been presented in support of the social historian's position. See, for example, Shapin (1982). In this paper, I concentrate on studies which deal with social factors related to society at large as opposed to the social structure of science itself for two reasons: (1) cases which involve "wider" social factors more strongly support the social historian's approach, and (2) the criticisms I develop apply with even more force to cases which involve social factors related only to the social structure of science.
11. Shapin (1975), p. 240.
12. This is a point that would be emphasized by social historians influenced by Mary Douglas's work (see, e.g., Barnes and Shapin (1977)). For what Douglas provides is an account of the way in which certain social structures give rise to the general categories of whole cosmologies. She has, for instance, suggested that isolated societies with strong boundaries between ranks and roles tend to adopt witchcraft cosmologies while societies with weak internal and external boundaries tend to favour individualistic cosmologies.
13. See, e.g., Young (1973), pp. 346-347.
14. For this criticism see, e.g., Ben-David (1971), p. 4.

15. Barnes (1982), p. 22. See also Barnes and Shapin (1979), p. 11.
16. This point is recognized by Steven Shapin, who observes that "neither Scrope's uptake of banking metaphors nor Darwin's uptake of Malthusian conceptions reveals anything in particular about the social purposes of the actors concerned". See Shapin (1982), p. 179.
17. Brown (1970), pp. 29-30.
18. See MacKenzie (1979), especially p. 126 and p. 132. A full study of MacKenzie's argument would involve a detailed examination of his conceptions of social interests and "exceptional individuals". These are central to his argument, but neither is unproblematic.
19. Hesse (1970), p. 138.
20. It should also be recalled that episodes involving "irrationality" can be readily accommodated within the intellectual historian's framework. Intellectual historians do not deny that social explanations of scientists' beliefs are sometimes required. Their claim is that as a matter of fact scientific beliefs rarely have social causes, not that they never do.
21. For this view see, e.g., Barnes and Shapin (1979), p. 9.
22. Bloor (1976), p. 13. See also Hesse (1978), pp. 59-60.
23. Barnes (1982), p. 30 and p. 32.
24. Hesse (1980), p. 37. See also Barnes and Bloor (forthcoming).
25. In this context it is worth recalling the point, stressed by William Whewell over a century ago, that the scientist's problem is normally not one of selecting from among the plethora of theories that cover the data but of finding "even one other to account for (them) besides the hypotheses by which they have been explained with such curious completeness" ((1860), p. 271). See also Lugg (1978).
26. Barnes (1982), p. 34.
27. Cf. Thomas Kuhn's observation that proponents of the "new internal historiography" set aside the science they know and treat their subjects in their own terms (Kuhn (1977), p. 110). And note that social historians almost always assume an exceptionally weak notion of

relativism. Barnes and Bloor, for instance, take it to comprise the following three remarkably uncontroversial theses: (1) beliefs about certain topics vary; (2) belief is context dependent; and (3) all beliefs without exception should be investigated empirically. See Barnes and Bloor (forthcoming), pp. 2-3.

28. Bloor (1976), p. ix and Mendelsohn (1977), pp. 2-3.
29. Wrong (1961).
30. For this view see Brown (1974), p. 185; Barnes and Shapin (1979), pp. 10-11; MacKenzie and Barnes (1979), pp. 191-192; and Barnes (1982), p. 39.
31. See Westfall (1971), p. 2 and McMullin (1979), p. 23.
32. Hesse (1970), p. 135.
33. Lakatos (1978), p. 102 and Laudan (1977), p. 202.
34. This type of explanation is sympathetically examined by Morton White in his (1965), pp. 194-200. Hesse has suggested that White's conception be applied to the history of science. See her (1973), p. 134.
35. For a detailed discussion of "rational explanation" along these lines see Aronovitch (1978).
36. This view is suggested by Shapin and Barnes's (1979), pp. 135-137. In their opinion, "it is regrettable that descriptions of the psyche are so critically and routinely accredited by historians" (p. 135).
37. The flexibility of the intellectual historian's approach can be appreciated by considering the acceptance by many Soviet biologists in the 1940s and 1950s of Lysenko's view that heredity is determined not by genes but by an internal property of living matter. Intellectual historians will not assume in advance that their adoption of this view must be explained in any particular way. They will attempt to determine whether the biologists came to their view because they accepted certain findings about, say, spring crops as conclusive evidence, or because they thought that persons in power were in a position to know, or because accepting the theory was the politically expedient thing to do.
38. For a recent discussion of the importance of considering the pragmatic aspects of explanation see Garfinkel (1981), especially chapter 1.

39. Cf. Young (1969), p. 132.
40. Compare Steven Shapin's discussion of Scrope and Darwin, in which he points out that "the availability and comprehensibility of given cultural items (varies) for different groups differently situated in the social structure and at different places and times" (Shapin (1982), p. 179).
41. The difference between explaining a contrast and explaining what it is that is being contrasted is discussed in detail in Garfinkel (1981). See especially p. 89.
42. It is worth noting here that differences in belief may sometimes be explained in terms of the locations of scientists in the social structure of science itself. Consider the debate that occurred in the 1860s between Pasteur and Pouchet concerning spontaneous generation. In a controversial paper, John Farley and Gerald Geison argue that the debate turned on other than purely scientific considerations. (See Farley and Geison (1970).) A less extravagant interpretation, however, can be developed by noting that Pasteur, whose early work was in crystallography, and Pouchet, who was a traditional biologist, were associated with different research traditions. We can explain their views by noting that science incorporates a vast division of labour and by indicating how the locations of the two scientists delimited the information to which they had access. For further discussion of this type of case see Lugg (1980), especially section III.
43. Ryan (1981). Ryan's discussion of the differences between what he calls "genetic explanations" and "circumstance explanations" is also relevant in the present context. See especially pp. 34-37.
44. Shapin (1982), p. 195.
45. See footnote 42. Like Shapin, I hold that there are no "interpretive and methodological asymmetries between sociology and history of science, or between the study of modern professionalized science and that of past settings" (Shapin (1982), p. 195). But I suspect that our reasons for holding this view are very different.
46. Lest it be thought that the concessions I have been

making to the social historian are insignificant, it is perhaps worth observing that considerations of the sort I have been discussing can be discerned in the work of some of the founders of the sociology of knowledge. In particular, I believe that Karl Marx's view in 'The Eighteenth Brumaire' and other works and Karl Mannheim's discussion in *Ideology and Utopia* are best interpreted within the general framework I have been sketching. See Marx (1869) and Mannheim (1936), chapter V. On Marx's views, see also Ryan (1981), p. 35.

47. The first draft of this paper was completed while I was visiting the Center for Philosophy of Science at the University of Pittsburgh in 1979-1980. I have benefited from discussions with members of the Center -- especially Larry Laudan -- as well as from the comments of Toby Appel, Hilliard Aronovitch, Lynne Cohen and Steven Shapin.

References

Aronovitch, H.: 1978, 'Social Explanation and Rational Motivation', *American Philosophical Quarterly* 15, 197-204.

Barnes, B.: 1982, *T.S. Kuhn and Social Science,* Columbia University Press, New York.

Barnes, B. and D. Bloor: forthcoming, 'Relativism, Rationalism and the Sociology of Knowledge'.

Barnes, B. and S. Shapin: 1977, 'Where is the Edge of Objectivity?', *British Journal for the History of Science* 7, 61-66.

Barnes, B. and S. Shapin (eds.): 1979, *Natural Order,* Sage, London.

Ben-David, J.: 1971, *The Scientist's Role in Society,* Prentice-Hall, Englewood Cliffs.

Bloor, D.: 1976, *Knowledge and Social Imagery,* Routledge and Kegan Paul, London.

Brown, T.: 1970, 'The College of Physicians and the Acceptance of Iatromechanism in England, 1665-1695', *Bulletin of the History of Medicine* 44, 12-30.

Brown, T.: 1974, 'From Mechanism to Vitalism in Eighteenth Century English Physiology', *Journal of the History of Biology*

7, 179-216.

Clagett, M.: 1959, *The Science of Mechanics in the Middle Ages,* University of Wisconsin Press, Madison.

Cohen, I.B.,: 1980, *The Newtonian Revolution,* Cambridge University Press, Cambridge.

Farley, J. and G. Geison: 1970, 'Science, Politics and Spontaneous Generation in Nineteenth-Century France: The Pasteur-Pouchet Debate', *Bulletin of the History of Medicine* 48, 161-198.

Forman, P.: 1971, 'Weimar Culture, Causality and Quantum Theory, 1918-1927: Adaptation by German Physicists and Mathematicians to a Hostile Intellectual Environment', in R. McCormmach (ed.), *Historical Studies in the Physical Sciences,* Vol. 3, University of Pennsylvania Press, Philadelphia, pp. 1-115.

Garfinkel, A.: 1981, *Forms of Explanation,* Yale University Press, New Haven.

Guerlac, H.: 1975, *Antoine-Laurent Lavoisier, Chemist and Revolutionary,* Scribner's Sons, New York.

Hesse, M.: 1970, 'Hermeticism and Historiography: An Apology for Internal History of Science', in H. Stuewer (ed.), *Minnesota Studies in the Philosophy of Science,* Vol. 5, University of Minnestoa Press, Minneapolis, pp. 134-160.

Hesse, M.: 1973, 'Reasons and Evaluation in the History of Science', in M. Teich and R. Young (eds.), *Changing Perspectives in the History of Science,* Heinemann, London, pp. 127-147.

Hesse, M.: 1978, 'Internal and External Causation of Scientific Ideas', in E. Forbes (ed.), *Human Implications of Scientific Advance,* Edinburgh University Press, Edinburgh, pp. 59-63.

Hesse, M.: 1980, *Revolutions and Reconstructions in the Philosophy of Science,* Harvester, Brighton.

Koyre, A.: 1965, *Newtonian Studies,* Harvard University Press, Cambridge.

Kuhn, T.S.: 1977, *The Essential Tension,* University of Chicago Press, Chicago.

Lakatos, I.: 1978, *Philosophical Papers,* Vol. 1, Cambridge University Press, Cambridge.

Laudan, L.: 1977, *Progress and Its Problems,* University of California Press, Berkeley.

Lugg, A.M.: 1978, 'Overdetermined Problems in Science', *Studies in the History and Philosophy of Science* 9, 1-18.

Lugg, A.M.: 1980, 'Theory Choice and Resistance to Change', *Philosophy of Science* 47, 227-243.

MacKenzie, D.: 1979, 'Karl Pearson and the Professional Middle Class', *Annals of Science* 36, 125-143.

MacKenzie, D. and B. Barnes: 1979, 'Scientific Judgment: The Biometry-Mendelism Controversy', in B. Barnes and S. Shapin (eds.), *Natural Order*, Sage, London, pp. 191-210.

McMullin, E.: 1970, 'The History and Philosophy of Science: A Taxonomy', in H. Stuewer (ed.), *Minnesota Studies in the Philosophy of Science*, Vol. 5, University of Minnesota Press, Minneapolis, pp. 12-67.

Mannheim, K.: 1936, *Ideology and Utopia*, Harcourt Brace and World, New York.

Marx, K.: 1869, *The Eighteenth Brumaire of Louis Bonaparte*, reprinted in L. Feur (ed.), *Marx and Engels: Basic Writings on Politics and Philosophy*, Doubleday, New York, 1959.

Mendelsohn, E.: 1977, 'The Social Construction of Scientific Knowledge', in E. Mendelsohn, P. Weingart and R. Whitley (eds.), *The Social Production of Scientific Knowledge*, Reidel, Dordrecht.

Mulkay, M.: 1979, *The Sociology of Scientific Knowledge*, George Allen and Unwin, London.

Rudwick, M.: 1974, 'Poulett Scrope on the Volcanoes of Auvergne: Lyellian Time and Political Economy', *British Journal for the History of Science* 7, 205-242.

Ryan, C.: 1981, 'Beyond Beliefs', *American Philosophical Quarterly* 18, 33-41.

Shapin, S.: 1975, 'Phrenological Knowledge and the Social Structure of Early Nineteenth-Century Edinburgh', *Annals of Science* 32, 219-243.

Shapin, S.: 1979, 'The Politics of Observation: Cerebral Anatomy and Social Interests in the Edinburgh Phrenology Disputes', in R. Wallis (ed.), *On the Margins of Science*, Sociological Review Monographs, Keele, pp. 139-178.

Shapin, S.: 1982, 'History of Science and Its Sociological Reconstructions', *History of Science* 20, 157-211.

Shapin, S. and B. Barnes: 1979, 'Darwin and Social Darwinism: Purity and Danger', in B. Barnes and S. Shapin (eds.), *Natural*

Order, Sage, London, pp. 125-142.

Westfall, R.: 1971, *The Construction of Modern Science*, Cambridge University Press, Cambridge.

Whewell, W.: 1860, *On the Philosophy of Discovery*, reprinted by Burt Franklin, New York, 1971.

White, M.: 1965, *Foundations of Historical Knowledge*, Harper and Row, New York.

Wrong, D.: 1961, 'The Oversocialized Conception of Man in Modern Sociology', *American Sociological Review* 26, 183-193.

Young, R.: 1969, 'Malthus and the Evolutionists: The Common Context of Biological and Social Theory', *Past and Present* 43, 109-145.

Young, R.: 1971, 'Darwin's Metaphor: Does Nature Select?', *The Monist* 55, 442-503.

Young, R.: 1973, 'The Historiographical and Ideological Contexts of the Nineteenth-Century Debate on man's Place in Nature', in M. Teich and R. Young (eds.), *Changing Perspectives in the History of Science*, Heinemann, London, pp. 344-438.

Robert E. Butts

THE ROLE OF ARATIONAL FACTORS IN INTERPRETIVE HISTORY: THE CASE OF KANT AND ESP

The Challenge of the Strong Programme. The strong programme in sociology of knowledge is a *programme;* it is a format for research and a philosophical manifesto. If it did not provide a reasoned context for doing research it would be of no interest to sociologists; if it professed no epistemology it would be of little interest to philosophers. Philosophers of science (in growing numbers, it seems) find a challenge in the tenets of the strong programme. That challenge has largely to do with the fact that the strong programme offers an alternative epistemology in the name of scientific sociology, and many philosophers find it an alternative sadly lacking in philosophical justification.

Much attention has been paid to the strong programme's symmetry principle, and to the problems thought to arise from applying that principle to causes or conditions of scientists' beliefs that are not rational. The symmetry principle flies in the face of the Mertonian arationality assumption discussed sympathetically and at length by Laudan in *Progress and Its Problems*. The principle challenges, among much else, the longstanding philosophical contention that reasons can be causes, and that in seeking to explain choices of beliefs, we should look for reasons for those beliefs, arguments in favor of those beliefs, before we give up the game and turn to social conditions -- or worse, psychological ones -- as explanations. In our heart of hearts we all know that social and psychological conditioning has an important effect upon what we believe, what we conclude, and what we accept. Feyerabend and Kuhn would not have such a profound hold upon us if we did not know this. But philosophers are committed to the inherent or consequential rationality of science (and philosophy as well), and

J. R. Brown (ed.), Scientific Rationality: The Sociological Turn, 209–247.

so for many of us acceptance of the symmetry principle is impossible. It may be that Leibniz accepted the calculus as a reliable mathematical instrument because of his gnostic upbringing, but this does not explain why the calculus was accepted by Leibniz as good science; to explain this we need to find the reasons for Leibniz' acceptance, the good arguments that he thought could be put forward on behalf of that acceptance.

Philosophers' reluctance to accept the strong programme, and especially its symmetry principle, seems to me to stem in part from philosophers' deep commitment to the proposition that science/philosophy really is the paradigm of rational activity, and to the following propositions that this commitment implies:

1. The epistemology of Western Science is the preferred epistemology.
2. The methodology of Western Science is the preferred methodology.
3. The rationality of Western Science is the only form of rationality that requires philosophical justification, and it is the only form of rationality that can be so justified.
4. Western Science is essentially philosophical: it investigates on the basis of shared, public methods; in so far as it is possible it infers canonically; it concludes on the basis of reasoned argument; it accepts on the basis of reasoned faith (in induction).
5. However one makes out the case, Western Science is progressive, it is cognitively successful -- it works.

These commitments as stated will not stand unchallenged. All of us would at once rush in with qualifications, nice distinctions, editorial suggestions -- as academics are accustomed to do, we would form a committee and take votes. But that won't do. Science and philosophy ought not to be done by committees. Taking a vote is an abdication of the responsibility to undertake the hard work of reaching a universally acceptable rational conclusion. Science/philosophy does not give notices of motion, it issues communications, publishes results. The communications may fail to communicate, and the results may be wrong, but we need to find this out, and to make suitable corrections, *by means of science*, by means of rational argument, including experimentation and perfection of instrumentation.

Advocates of the strong programme can probably accept all that I have said so far. Accepting my characterization of the commitments of a rational Western Man, they would simply go on to point out that all they seek is freedom (from philosophers' carping) to engage in the kind of science we admire. Essentially what they seek is license to investigate science itself scientifically. They know (at least Bloor and Barnes seem to know) that science is a body of knowledge as well as a set of social and political activities, and for this reason they also know that studying science scientifically involves appeal to many sciences, among them sociology and psychology. If sociology and psychology of science were content to limit investigation to investigation of what we might call the cultural background of scientific activity, the external social and psychological trappings of scientific work, there is no negative philosophical response that could make sense.

However, the strong programme is not an extension of Mertonian diffidence. Instead, it promises to investigate *the internal theoretical workings* of science scientifically (sociologically). To do this it will abrogate the arationality assumption and seek direct sociological explanations of why scientists accept the results that they do. To many philosophers, this means that explanations of scientific inferring and concluding will *effectively bypass* two central disciplines: theory of rationality (taken to be philosophically essential in characterizing the nature of scientific theory); and intellectual history in the form of history of science (taken to be the central ancillary discipline of philosophy of science, providing, as it does, the data of science as a cognitive activity). It is this threatened bypass operation that some philosophers perceive to be the unkind cut.

Philosophers are not troubled by sociologists who study scientists as members of instituional groups, as people who testify before appropriation committees, who perform on television. It is only when it is alleged that these activities of scientists have any bearing on the cogency of their theoretical or experimental results that philosophers bestir themselves. As Laudan has amply shown, it is the status of *cognitive* sociology that is at issue, it is the sociological study of scientists' beliefs.

Arational Factors and Utilities in Historical Explanation*. I have myself engaged in studies in interpretive history of ideas, and often my interests have led me to what some regard as the seamier side of scientific or philosophical belief. For example, without hesitation I attributed Leibniz' acceptance of the strange theory of the monads to his gnostic and mystical background, a background that most of us would identify as irrational, or certainly arational. The preferred mode of studying Leibniz or any other figure in the history of science/philosophy, at least amongst historians of philosophy, involves bringing the thought of Leibniz up to today's standards of rationality. The point of the exercise is to show how much in the thought of Leibniz is of lasting value because it maps onto a present presumed-to-be-correct theory of science. I have no objection to studies in this mode. They are important and illuminating. Using our present understanding of logic to help capture what Leibniz was doing when he invented the prototype of our logic is indispensable. Using our present understanding of physics to help capture Leibniz' concept of living force is likewise indispensable. To dispense with current forms of rational paradigms is much like dispensing with the activity of intellectual history altogether. For, among other things, that history has responsibility for giving an account of the continuities in scientific thought. And we cannot know why we are where we are unless we relate where we are to where others have been.

But there is more to consider. Leibniz did write the German documents, was for a time secretary of a Rosicrucian lodge, did talk about angels. Strong programme advocates would, I suppose, want to make these features of Leibniz' activities part of the explanation -- the sociological explanation -- of his scientific beliefs. At the very least they would want to hold that explanation of these beliefs is not different in kind from explanation of Leibniz' "correct" scientific beliefs. Laudan, I suppose, would want to say that they are a part of the metaphysical constitution of a research tradition, and that in the long run they turned out to be nonprogressive features of that tradition. Certainly they do not constitute reasons for Leibniz' acceptance of his progressive scientific beliefs. But on neither of these views is the seamy side of Leibniz eliminated. It should not be eliminated. It is there for all to see. It must be dealt with. The only question is how? Allow me to move to another example in an attmept to show the role I think the intellectual historian can play in clarifying the issues, if not in resolving the dispute. The scheduled bypass surgery must alarm the historian of ideas as much as it alarms the philosopher.

Before I discuss the example let me anticipate the moral of my story -- I mean to dull the edge of the scalpel before it cuts very deep. I believe that what are sometimes taken to be social and other external factors do play a role in explaining scientific beliefs, but that this role is not performed by these often arational factors taken as universal ingredient in explanations, or as instances of such generalizations. Scientists undertake their studies for reasons that are better seen as expected ultilities. Scientific explaining is, as is often remarked, goal directed. A scientist does his work with certain ends in view (as Dewey would put it), and often, if not always, he will accept the results of a given piece of work because he also believes that the work has achieved those ends in view. Maybe all he seeks is the truth, but epistemic utilities are still utilities.

Accounting for satisfaction of goals as reasons for accepting scientific results is usually done in ways that see utilities as internal features of scientific induction. For Bayesians, the utilities may be expressed as prior probabilities and may refer informally to expectations of success based on kinds of past success. Wesley Salmon has suggested that the history of science functions in Bayesian inference in something like this way in his provocative "Bayes's Theorem and the History of Science."[1] I will seek to show quite informally that intellectual history appeals to expected fulfillments of ends in view as explanaations of accepted beliefs in something like this Bayesian way.

What I am suggesting is this. One form of explanation (Gutting's Type II explanation) has a sociological generalization as its major premise:

(i) Anyone in circumstance C believes p. (*Sociological generalization*).
(ii) S is in circumstance C.
(iii) Therefore, S believes p.

The form of explanation I am now advancing replaces the sociological generalization with an expressed or implict end in view, a *meta-reason* for undertaking the research in the first place. This gives us a modification of Gutting's Type I explanation:

(i) S believes p'.
(ii) S reasons in manner R (*for meta-reason M*) and believes the results of his reasoning.
(iii) Given p', reasoning in manner R (*for meta-reason M*) results in p.
(iv) Therefore, S believes p.

I am aware that meta-reasoned reasoning is often self-serving, and that its employment can validate deductively formulated explanations in unacceptably circular ways. (A reasoner's meta-reason for accepting reasoning in manner R might be that he will believe anything than follows in manner R from p'.) My suggestion is simply this: sometimes, at least, a reason for engaging in research of a given kind (and employing a methodology of a given kind) is not a social or psychological or other form of *external cause,* but is a reason as an end in view (maybe only the end of discovering the truth), that is effectively internal to the task of research. When the intellectual historian lights upon such a meta-reason, as he often does, he has discovered an important and relevant *internal reason* for the scientist's belief. And this is true even if the meta-reason is, as it often is, an *areason,* that is, something that the scientist accepts for *whatever reason.*[2]

Much of what is involved in the suggestions I am making has to do, I think, with one of the deep problems inhering in the strong programme, and one not often enough remarked. Advocates of the strong programme have the wrong conception of philosophy! I will not push the point, but it does seem to me that Bloor accepts as his model of philosophizing just that kind of sleight of hand informal analysis of language that is most distant from the work of scientific reasoning. I think one can agree that if philosophy is seen as the kind of thing done by Austin and Flew, then scientists, sociologists included, should put as much distance between themselves and philosophy as possible. But there is another kind of philosophy, one that has as ancestors those who would not have recognized the difference between philosophy and science -- Descartes, Newton, Leibniz, Kant -- and as more recent practitioners Whewell, Mill, Peirce, Dewey. This kind of philosophy -- call it what you will -- sees philosophy as at least continuous with science, and it is the view of philosophy as basically not distinct from science that underwrites the kind of analysis I am giving here.

A Case Study: Kant and the Paranormal. Now my case study in the history of ideas, one that I hope will confirm the results of the analysis given above, and will seek again to emphasize Laudan's surely correct observation that *"the application of cognitive sociology to historical cases must await the prior results of the application of the methods of intellectual history to those cases."*[3]

My example is the case of Immanuel Kant's attitudes toward the doctrines of Immanuel Swedenborg and in general toward alleged cases of extra-sensory perception. The case should interest historians of science because in discussing Swedenborg Kant introduces an encapsulated theory of matter, remarks upon a solution to the mind/body problem, including intriguing references to physiological explanation of mental events, and continues the development of the "double government" methodology of Leibniz. Philosophers should be interested in the case because of the connection between Kant's discussion of Swedenborg and Kant's own later ethical theory. Up until now, there has been little interest shown by either philosophers or historians of science in Kant's views on ESP, except by those who want to show that Kant's references to a spiritual world entail a theory of spirits that is derived, but not acknowledged by Kant to have been so derived, from Swedenborg.[4]

In 1766 Kant published what is generally regarded by Kant scholars as his most remarkable, because atypical, book: *Träume eines Geistersehers erläutert durch die Träume der Metaphysik.* The '*Geisterseher*' is Swedenborg. Unlike all of Kant's other works, this book is written in a partly playful style; it teases and insinuates; it is contemptuous. C.D, Broad thought that it abounds "in elephantine badinage." Broad thought -- incorrectly -- that there is only one other place where Kant discusses or refers to Swedenborg, the letter to Miss Charlotte von Knobloch of 1763.[5] We now know that Kant made a regular practice of referring to Swedenborg in lectures on rational psychology. There are also references in two later works, and two Reflexionen.[6] I have also been able to find three more letters, two to Moses Mendelssohn dealing with *Träume,*[7] and a late (1790) letter to Borowski, in which Kant deals with the *Schwärmerei,* the enthusiastic or ecstatic dream state, brought about by too much reading.[8]

Kant's letter to Charlotte von Knobloch is evidently a reply to a letter from this lady to Kant.[9] Kant details three now well-known cases in which it had been reported that Swedenborg experienced supernormal congition or achieved extrasensory perception. Kant had learned -- the details need not trouble us here -- that Swedenborg was reputed to have had a clairvoyant conversation with the dead brother of the Swedish queen, during which converstion Swedenborg learned a secret known only to the queen; that Swedenborg had "discovered" a lost receipt for payment to a goldsmith in the secret compartment of a dead man's bureau; and that Swedenborg, while in Göteborg -- 459 kilometers from Stockholm -- had "seen" a fire that was stopped just three doors from his own house. Kant had sent a letter to Swedenborg inquiring about these alleged incidents which evidently Swedenborg received. But he did not reply. One of Kant's English merchant friends had been dispatched to discuss the incidents with Swedenborg, which he dutifully did. He reported that Swedenborg had received Kant's letter, would reply in a book soon to appear, and that the best testimony the Englishman could find attested to the truth of the stories reported to Kant. Kant adds that he has received other stories of Swedenborg's extensive communications with spirits in the supersensory world.

Kant begins his letter to Charlotte von Knobloch by expressing charitable agnosticism concerning the reported feats of mystical communciation attributed to Swedenborg. He says that although we cannot deny the possiiblity of such events, we must investigate carefully, for we know little about the nature of spirits, if indeed there are such. In the past,he was inclined not to accept the credentials of such alleged happenings. However, he expresses a strong interest in the Swedenborg cases, and obviously went to some trouble to verify them as soundly as he could. He also expressed an interest in interrogating Swedenborg himself on these matters. So we know that Kant was fascinated by the reports, sought more information, and reserved judgment, as indeed we would expect from a circumspect reasoner like Kant.[10]

The letter to Miss von Knobloch would lead most of us to think of Kant, the rational scientist, as taking an interst in any fascinating event that requires a difficult explanation. He had already shown that he was not beyond speculation of a marvelously imaginative sort. Thus in his 1755 treatise on the nebular hypothesis, the *Allgemeine naturgeschichte und Theorie des Himmels, oder Versuch von der Verfassung und dem mechanischen Urspringe des ganzen Weltgebäudes nach Newtonischen Grundsätzen abgehandelt,* we find references to "various classes of intelligent beings," "kinds of thinking natures," and to the various "dwelling-places" of these "intelligent creatures." He writes about the dependence of the "spriritual faculties" of the various inhabitants of planets on grosser or finer, heavier or lighter, matter as determined by the "distance of these abodes from the sun." The inhabitants of Jupiter or Saturn belong to the "most exalted class of intelligent creatures. These at least have a different Time-idea from ours; they are not subject to death in the same degree that we are..." Man occupies a middle ground between these most excellent and the more imperfect grades of "thinking natures."[11] We know that in the middle decades of the century Kant was still struggling with the question of a proper method for metaphysics, and that he had not yet freed himself from the spell of Leibniz. Although Leibniz had insisted that invesgation of nature should proceed mechanically, he also believed that one could account for the nature of things according to principles of perfect reason, principles that taught us that reality is a continuity of nonextended simple substances that are internally teleologically oriented and organized. The "double government" methodology of Leibniz was obviously still in Kant's mind during this period, and indeed it never left his thought throughout his entire career. The speculations in the third part of *Allgemeine Naturgeschichte* are typically Leibnizian [12]

Kant and the Double Government Methodology of Leibniz. I want to dwell on this Leibnizian background of Kant's thought for a moment, because I believe that Kant never abandoned Leibniz' intent to harmonize teleological and mechanical forms of explanation, and because I think Kant's acceptance of the "double government" methodology provides the clue to an understanding of his apparently changing attitudes to Swedenborg. As we will see, this is a clue that Broad did not pick up in his own attempt to understand these attitudes. Kant's acceptance of this methodology also provides us with access to a prominent feature of his meta-reason, his arational motivation, for concluding that Swedenborg's ideas must be taken very seriously, so seriously as to be worth protecting from frivolous distortion in the public mind.

In many places in his writings Leibniz undertakes to develop and defend what he refers to as the mechanical method. Briefly, this is a ratiocinative and experimental strategy for explaining the behaviour of bodies. Physical phenomena are to be reduced to forms of expression in terms of magnitude, figure and motion. The model of physical interaction is Cartesian; things collide with one another. Leibniz recognizes, however, that since for him bodies are not quite real, this mechanical method will not serve to explain real substances (the monads), which do not collide with one another, and must be understood in terms of another model: that of perception and appetition. This metaphysical method construes substances teleologically, rather than with reference to efficient causes. Mechanical causes are replaced by reasons.[13]

Leibniz gives a rather full statement of what is involved in this methodological dualism in *Critical Remarks Concerning the General part of Descartes' 'Principles'*.[14] Leibniz challenges Descartes' claim to the exclusivity of the mechanical method as the method of adequate explanation, and urges inclusion of teleological factors in metaphysical explanation. Leibniz accepts that the mechanical method is fully adequate to explain the causes of behaviours of material objects, and that physics requires no access to occult qualities or supersensible causes. Nevertheless, causes of material object transactions are themselves derived from "higher" metaphysical causes, to which appeal must be made to sustain complete explanations of the nature of things. This explanatory dualism leads Leibniz to this extraordinary conclusion:

> Nature contains, so to speak, an empire within an empire, or a double government: the government of reason and the government of necessity, or the empire of forms and that of material particles.... Thus, if you study the necessary connection of material phenomena and the order of efficient causes, you will find that nothing happens without a cause that satisfies the imagination and nothing escapes the mathematical laws of mechanism. If, on the other hand, you contemplate the golden chain of ends and the sphere of forms which constitute, as it were, an intelligible world, you will recognize that, due to the perfection of the supreme author, the apexes of ethics and of metaphysics coincide so that nothing happens without supreme reason.

Leibniz' attempt to harmonize both mechanical and teleological forms of explanation by appeal to a rational metaphysics is mirrored in Kant's critical attempt to foster an epistemology and a methodology that also does justice to teleology in the face of a preferred mechanism for explaining the physical world.

Thus the Double Government Methodology (DGM) is also the methodology of Kant, and of Swedenborg (and of most of the rest of us). I need to show in some greater detail the depth of Kant's commitment to DGM. Perhaps it is best to ask: just what parts of the methodology of Leibniz did Kant accept? 1) DGM commits Leibniz to acceptance of mechanical explanation as the preferred form of explanation of the behaviour of material objects. Kant accepts this. 2) That form of explanation argues for the elimination of occult qualities and the abandonment of theological appeals in science. Kant accepts this. 3) DGM introduces the duality of reason and natural necessity. The dualism will be greatly transformed by Kant, but he accepts it nonetheless. 4) DGM provides for the separation of science and morality, and identifies the ethical and the metaphysical. Kant's mature works endorse the separtion and seek to justify the identification. That his early orientation inclined him to accept this methodology does not surprise us. Kant's Wolffian dependence upon Leibniz is quite obvious in the early works. But we continue to find the impulse to accommodate both mechanism and teleology in the *Träume,* in the *Inaugural Dissertaion* of 1770, in the great 1st *Critique*, in the mature ethical treatises, and most cflearly articulated in the *Critique of Judgement* of 1790, the work in which Kant finally confronts the threatened antinomial clash between mechanism and purposiveness, and works through to his form of DGM based on regulative employments of explanatory maxims.[15] For present purposes it is not required that I trace Kant's career-long struggles to ground DGM, a grounding finally achieved within the framework of the critical philosophy. Those who would show that Kant is heavily indebted to Swedenborg normally stress the perceived elements of Swedenborg's thought occurring in *Traume,* the *Inaugural Dissertation,* and the 1st *Critique*. I will have something to say about Kant's use of DGM in these works, but I want to return now to a resetting of the problem in the context of the present examination of Broad's problems with Kant's attitudes toward Swedenborg and supernormal cognition.

From this point on, then, I will take it as given that Kant accepts as an established meta-reason for investigating the problems he does, in the forms in which he does, some strong form of DGM. I will return to further details below.

Dreams of a Spiritseer. I turn then to a discussion of the *Träume*. The work is in two parts. Part I investigates the metaphysics of disembodied spirits and contains both a remarkable theory of matter and a contemptuous examination of the claims of those who profess communication with inhabitants of the spirit world.[16] Part II is in three sections, the first of which deals with the three stories about Swedenborg that Kant had written about in his earlier letter to Miss von Knobloch. The second section discusses the doctrines of Swedenborg that are apparently based on his personal conversations with spirits and his extrasensory perception. Kant's account is remarkably full and clear, especially for one based on the *Arcana*. Kant had been disappointed that he received this work, rather than an expected different one dealing directly with theoretical features of Swedenborg's observations. The final section of the book deals briefly with the utility of undertaking serious study of supernormal cognition. In short, Kant suggests that we leave ESP and related paranormality to the dreamers, and get on with the serious business of life.

Despite its playful tone and its unaccepting attitude toward Swedenborg's *Schwärmerei*, the book is a remarkably faithful statement of Leibniz's DGM adapted to a metaphysics in which man is seen as a member of two worlds, one a world of body, the other, of things of the mind, of which ethical pursuits are for Kant here, as elsewhere and subsequently, uppermost. He is careful to point out that the problem of the existence of spirits -- the metaphysical problem -- is recalcitrant to direct management by means of DGM. This is because everyone seems normally doubtful about the existence of spirits; thus the existence of spirits is not suggested by experience. Rather, "... many conceptions arise in secret and obscure conclusions incidental to experience, and afterwards are transmitted to other minds without even the consciousness of that experience or conclusion which has first established the conception. Such conceptions may be called 'surreptitious.'" This concept of a surreptitious cognition is the prototype of what Kant will call a 'subreptic axiom' in the *Inaugural Dissertation*, an axiom derived by misusing a sensitive concept as an 'intellectual phenomenon.' Of course for Kant the latter phrase is a contradiction in terms.[17]

Apart from its nonstandard and unscholarly style, and the fact that it was published anonymously, what is there about *Träume* that makes many question why Kant troubled to write it? It seems to deal with especially troublesome and "unscientific" matters. It is suggested that in Kant's time and place taking paranormal phenomena seriously is equivalent to experimenting with LSD at Harvard.[18] Kant was not yet a professor in the Albertus University of Königsberg, but only a *privat dozent*. Thus, quite apart from the fact that the little book contains a compact form of views that Kant will later give expression to in "scientific" works, it seems reasonable to wonder why the untenured teacher would risk all in writing it.

We thus have two questions to confront: why did Kant write the book at all? and why did he publish it anonymously? My answer to the first is that it is a book on metaphysics embodying Kant's continuing interest in locating the methodological limits of empirical knowledge. Given *that* context, Swedenborg's claims to ESP, precisely because they were made by a well-connected scientist who was himself of widespread repute, were exactly the kinds of claims to take seriously as test cases of Kant's sophisticated recasting of the Double Government methodology. My answer to the second question is that although the book appeared anonymously, Kant took special pains to get it into the hands of just those official and semiofficial persons who could have done him the most harm, and not to try to hide authorship of the book from them. These answers lead me to propose a modified Gutting Type I explanation for Kant's interst in the paranormal. But first, some consideration of Broad, who proposes a Gutting Type II explanation in precisely the kinds of terms that are grist for the mill of the advocates of the strong programme.

C.D. Broad's Sociological Explanation. Instead of a volume by Swedenborg presenting and arguing for his claims to seership, Kant received the eight volumes of the *Arcana Coelestia*. Having waded through these ravings of a "transcendental epileptic," a bored and offended Kant struck back. Now Broad says that this explanation is offered "very tentatively." Broad knew what is in the book, an examination of the limits of empirical cognition. Nevertheless, his suggested psychological explanation, though very tentative, must be taken seriously, because we are trying to account for adoption of a literary style largely foreign to Kant. Kant wanted to prevent as many as possible from taking too seriously the substantive claims of Swedenborg, who Kant thought had mislocated and misdescribed the intellectual world.

It is Broad's extraordinary sociological explanation of the need for anonymity that mainly distresses me. This explanation is as follows: "It is reasonable to suspect that, in 'enlightened' academic circles in East Prussia in the middle of the eighteenth century, a reputation for having carefully read Swedenborg's writings and having paid serious attention to the evidence for his alleged feats of clairvoyance, would be enough to condemn a *privat dozent* to remain in that position for the rest of his life.... I can only suggest that the conventions of the time and place permitted a *privat dozent* to flirt with this disreputable subject, provided that he made an honest man of himself by maintaining the *form* of anonymity and by adopting a sufficiently bantering and condescending tone towards the alleged phenomena and the persons of whom they were narrated. If these were the conditions, Kant certainly complied with them."[19] Now that is as good a Gutting Type II explanation as one could wish. It is, however, false.

The case for Broad's sociological explanation is considerably weakened by imperfections in his intellectual history. He thought that Kant's references to Swedenborg were limited to the letter to Charlotte von Knobloch and the *Träume*. But there are other materials to which I have referred, including the two letters to Mendelssohn, and for reasons I will now detail these must be taken very seriously. In the letter of February 7, 1766, Kant says that he is sending Mendelssohn copies of "some 'Träumereien'", one for Mendelssohn himself, and for six others. He says of the work: "It is a hastily prepared treatise, and presents rather an outline of the manner in which questions of this kind may be treated than the treatment itself."

Kant repeatedly sought Mendelssohn's approbation of his works in metaphysics; there can be no doubt that he sought it in this case. What of the six others? Two are Court servants, Court Preacher Sack, and Councillor of the Consistory Spaulding. I have made no effort to trace these gentlemen, but it must be noted that Kant made no effort to hide his impious book from two persons obviously highly placed in religious officialdom. If anyone could threaten his tenure in the university, these two could. The other four persons are all members of the Berlin Academy and hence also royal appointees (Mendelssohn had been proposed for membership, only to be blackballed by the King himself). These gentlemen are Provost Süssmilch, and Professors Lambert, Sulzer, and Formey. Johann Heinrich Lambert was regarded by Kant as "the leading genius of German." He was the only member to be admitted to all four classes of the Berlin Academy: mathematics, natural philosophy, speculative philosophy, and letters. Kant had the highest regard for his work, and for his opinions. Johann Peter Sussmilch gained some repute for defense and development of Hamann's theory that nature is *logos,* believing that humans learned language by "divine instruction." He was the target of Herder's critique of the divine origin of language. Both Kant and Swedenborg thought of man's response to the supersensible as *symbolical,* as in some sense linguistic. Perhaps Kant thought that his little book made a modest contribution to the disputes about the origins of language.

Samuel Formey was the permanent secretary of the Berlin Academy, and a popular interpreter of the philosophy of Wolff. He contributed to the widespread feminine interest in Wolff's thought with his 6 volume *La belle Wolffian* (1741-53). The author and editor of nearly 600 books, he sought to cancel the boredom of his declining years by issuing his own *Oeuvres posthumes*! As permanent secretary of the Academy he wielded considerable influence, and was thus a figure Kant needed to take seriously. Johann Georg Sulzer, a Swiss, was a very prominent member of the Academy. With Euler, he had sponsored Lambert for membership in the Academy. In his own philosophy he sought to replace ideation with feeling in working out the development of the Wolffian philosophy, a move which probably impressed Kant favourably. He translated Hume into German. He created the essay competition of 1763, won by Mendelssohn, but resulting in the publication of Kant's second-place effort now known as the *Prize Essay*. Kant often sought his response to his work, a response that was frequently quite limp, as in the cae of his reaction to the *Inaugural Disseration*.

This must count as an impressive array of highly-placed individuals to whom to send copies of an anonymously authored book. If Kant had really intended to keep a secret of his authoriship of the *Träume,* he did nothing that I could learn to promote that secret. Instead, he seems actively to have sought reactions from prominent churchmen, academic politicians, and quite important philosophical scholars. Thus, however one decides the question of anonymity, I think we must conclude that Kant took the message of his book very seriously and wanted that book to be regarded as an important contribution to the studies he was making into the foundations of metaphysics. Further confirmation of this thesis is obtained from Kant's second letter to Mendelssohn of April 8, 1766, to which I now turn our attention.

The Second Letter to Mendelssohn. Kant begins by thanking Mendelssohn for forwarding the copies of the book. He next expresses regret that Mendelssohn was offended by the tone of the book, a tone that Mendelssohn found "between jest and earnest." I will discuss the rest of the letter in some detail, becasue in these comments to Mendelssohn Kant decisively makes my point that he was motivated to write the *Träume* by a cluster of meta-reasons stemming from his search for a correct form of the Double Goverment Methodology. If I am correct in this, Broad's Gutting Type II sociological explanation is not only wrong, but an impediment in the way of a correct understanding of Kant's interests in the paranormal.[20]

Kant's second letter to Mendelssohn seems to me to answer all of the remaining questions surrounding his book on Swedenborg. Clearly, he wrote the book with some reluctance, and in a state of confusion not characteristic of Kant. The actual production of the book seems to have been astonishingly casual, if we believe Kant that it was printed a proofpage at a time, so that one sheet would be off at the printer's while Kant worked on the next. Nevertheless, reluctant, confused, and casual as Kant's attitudes and actions may have been, it seems very clear that he thought the object of the exercise worthwhile, if not imperative. That object was to call into question both a method -- the "dream science" -- and the alleged cases of paranormal cognition. It is the "contagious," hence irrational, unhealthy, nature of the *Schwärmerei* that so distressed Kant, especially at a time when his own sceptical yet positive critique of metaphysics was in the early stages of development. In short, he took both Swedenborg's ponderous science of the supersensible and his reputation as a storied clairvoyant to be threats to the Double Government Methodology as a scientific manifesto and as ultimate protector of public morality. Kant's interests as revealed in the *Träume* are soteriological in a suitably secular sense.

Kant says in the letter that he is seeking a "specific

criterion," presumably one that will systematically rule out cases of supposed direct access to spirits. That criterion, and its attendant doctrine of method, will appear in mature dress in several forms in the 1st *Critique*. In general, it stems from Kant's critical insistence on separating questions of knowing (within the bounds of possible experience) from questions of pretended insight through intellectual intuition (surreptitious cognizing once again). The specific rational claims to direct insight into the soul are of course examined by Kant at length in the paralogisms section of the *Critique*, where the surreptitious pretense of rational psychology is fully exposed. Kant's methodological conclusions will be harsh: in the 1st and 3rd *Critiques* he will insist that the existence of spirits cannot be hypothesized, cannot even be held to be likely or probable (*or* unlikely or improbable), because spirits are *categorial impossibilities*, not knowable within the bounds of possible experience.[21]

Kant tells us in the *Träume* that the origin of the belief in spirits is surreptitious. We learn more on this topic from the letter to Mendelssohn. Here Kant says directly that the central problem is one of data or evidence that would give assurance of the way in which the soul is present in the world, and, most crucially, he takes this problem (of the union of soul and body) to be a sub-case of the general problem of any kinds of connection between external powers and faculties that are merely receptive. This, of course, is *the* problem of the *Critique of Pure Reason* -- we may fairly take it to be the single most important question Kant faced in his long career. In the letter he also mentions what will become the motivating theme of the Dialectic of Pure Reason: can inferences of reason (he will later add, "unaided by intuitions") discover "primitive powers," that is, real causal powers? The ultimate pre-critical *critical* note is sounded in the final quoted sentence of the letter: "Here we must decide whether there really are not limitations established by the bounds of our reason, or rather, the bounds of the experience that contains the *data* of our reason."

The question of data, the question of evidence -- the question that is the nerve of DGM: the divide that finally cannot be crossed, the limit of the knowable. The philosophical crucifix. Kant was putting Swedenborg to the philosophical experiment of the cross. That, as I read the evidence, is the message of the *Träume*. It is a message fully consistent with Kant's philosophical worries *throughout* his career. It is a message revealing the enthymematic premise that functions as the meta-reason for Kant's troubled attitudes toward Swedenborg and the paranormal. That Kant's style in dealing with spiritualism is partly playful and partly scornful seems to me to present no special problems. It is not easy for a serious philosopher to drive the nails into the hands of a distinguished and popular foreign colleague. The more telling point, however, is that the question of Kant's atypical style is seriously diminished in importance once we understand how central a role DGM plays in Kant's *Träume*.

As I understand these matters, the two letters to Mendelssohn that he apparently did not know about, coupled with the direct availability of employment of the Double Government Methodology in *Träume*, provide a decisive refutation of Broad's sociological/psychological account of Kant's interests in the paranormal. I also take it that I have made part of Laudan's point: the intellectual history, properly worked out, cancels the need for sociological intrusions into explanations of scientists' beliefs, in just those cases where arational factors are intermixed with, or presupposed by, the scientists' accepted forms of reasoning. The arational factors, thus understood, can be ingredients in a rational reconstruction of scientific acceptance. I take it to be the central task of the historian of ideas to locate such arational factors -- as I have tried to do in the present study of Kant -- wherever possible, and to give guidelines concerning the incorporation of these factors into a rational account of scientific belief. This is not to deny that sometimes (often?) arational -- even irrational -- factors play a role in the beliefs of scientists; it would be strange if this were not so. Nor is it to deny that sociological explanations of science as institutionalized behaviour is impossible. It is just to deny the primacy of sociological forms of explanation at every level. If my examination of the case of Kant can be generalized to other cases, as I believe it can, then internalist explanations of belief can often take the form of appeals to arational factors as meta-reasons within the larger context of reasoned cases of acceptance. Such factors tell us important things about what a given scientist takes to be at stake in pursuing the line that he does. And where, as in the present case of Kant, the utilities involved in a single piece of work find more extensive use and expression in other parts of the scientist's endeavors, the case for inclusion of arational commitments as meta-reasons in the reasoning becomes compelling.

Träume and the Inaugural Dissertation. To complete my study of the case of Kant and ESP, I have, then, one more task to perform. I must show, at least in outline, that indeed Kant's *Träume* commitment to DGM did continue to motivate his further researches. I have mentioned some connections between the problem of spirits that preoccupied Kant in *Träume* and his treatment of that problem in the *Critiques*. It will help to make my case for Kant's abiding commitment to DGM if I say a few words about continuation of his interests in the pre-critical period, interests revealed fully in the *Inaugural Dissertation*, and a few words about his positive development of interest in spirits given expression in his introduction of the idea of a *corpus mysticum* in the 1st *Critique*.

It has not been my aim in this study to expose the details of Kant's *Träume*, but I want now to say something more about his theoretical and especially methodological conclusions in this work. After showing that the presumed power of spirits to act on one another and on bodies cannot, like physical force, be located, Kant next develops a long story about what it would be like to be human and a member of two worlds, one physical and the other spiritual. Such a human "ghost-seer" would possess all manner of paranormal possibilities of cognition, would, in short, have ESP. The "communication" between seers could not be direct, because there is no intuitive sensory content in paranormal cognition; it would therefore be *symbolic*, largely a matter of interpreting signs given either as material objects or as words. The communion of spirits could not be located or intuitively represented except in quasi-sensory ways. Intelligibility in the world of spirits would be a matter of gnostic privilege in reading symbolic messages. A person like Swedenborg, who takes himself to be member of such a spirit world (while at the same time being a member of the world of bodies), is either a genuine spirit-seer, or is in a pathological state of delirium. Kant suggests that the two states are really indistinguishable, and that one who chooses to think of all alleged cases of ESP as pathological has certain advantages, for example, he can easily dismiss all allegations of the paranormal just because they fail to comply with ordinary demands of coherence in common experience.

Kant's theoretical conclusions are two: [1] the paranormal and the physiologically pathological are not clearly distinguishable; and [2] we are, and are doomed to remain, wholly ignorant of solutions to three basic problems: (i) the problem of the animation of a human body by a human soul at conception; (ii) that of the continuing connection between a soul and a body during life; and (iii) the separation of the soul from the body at death and its possible continuing post-death existence. His methodological conclusion points the way to further developments in the critical philosophy. To be genuine, a scientific hypothesis must deal with possible objects, and must speculate about detailed features of the behaviour of such objects. However, we can only know about possibility given actuality, and the only actual objects we have available to us are encountered in sensation. Possibility must have intuitive content (hence the need for Kant later to appeal to constructions in mathematics which give intuitive content). His methodological conclusion in *Träume* is therefore sceptical and agnostic, as he told Mendelssohn in the second letter it had had to be. Both the theoretical and methodological conclusions of this work are consistent with Kant's adoption of DGM. Science is limited in its investigations to the physical world where the connections between objects are mechanical and deterministic; the other "world," the world of cultural presence, of morality, art, and law, is a world that can only be understood by perfect reason -- which for Kant will finally mean *practical* reason.

I turn now to Kant's *Inaugural Dissertation*, defended in 1770, just four years after the appearance of *Träume*. I take it that the main themes of the dissertation were in Kant's mind when he wrote the second letter to Mendelssohn, that is, in the year of the appearance of *Träume*. It is of course now well established that the dissertation is the first of Kant's "critical" works, containing important new ideas the full working out of which would occupy Kant for the next decade. What is not well recognized is that the main lines of argument in the dissertation follow more or less exactly upon the conclusion of *Träume,* and amount to a first attempt on Kant's part to solve the general problem of any form of external impact on receptive faculties. He had told Mendelssohn, moreover, that the central question is one of data; in the dissertation, we find Kant's first appeal to the singularity and immediacy of data or evidence in the introduction of this special concept of an "intuition".

The *Inaugural Dissertation* begins by an investigation of the properties of "a world."[22] No one could accuse a young follower of Wolff of idle curiosity in further pursuit of understanding of which worlds are possible, which actual. But Kant's interests turn out to be rather special; his thought in the dissertation already takes the epistemological turn -- his deep question concerns which worlds are knowable. Kant begins the discussion in a typically scholastic way (remember, this work is one Kant had to defend publicly and show to be fully compatible with current modes of doing philosophy).[23]

Briefly reviewed, this is what the exercise of Kant reveals: *Analysis* arrives at a part which is not a whole, a *simple*. *Synthesis* arrives at a whole which is not a part, a *world*. The logically most primitive idea of a world (call it World I), is that of a composite, or set: from a numerically distinct collection of parts, the intellect forms the abstract notion of whole. There may be many such worlds; being creatures of the intellect, they are governed solely by the law of non-contradiction. Kant takes the problem to be one of representing this abstract notion concretely by means of a distinct *intuition*. Call this intuitively presented world World II. Such a world, Kant thinks, would be subject to certain temporal conditions, as well as logical ones, the temporal conditions giving grounds for *successive addition of parts to generate a whole*. Being subjected to time means being subjected to intuited singularities coming one after another; singulars, not wholes, are given. His question now becomes: how many worlds of World II type can there be? To answer this question we must look at two of the special kinds of properties he thought of as characterizing worlds.

The *matter* of a world is its parts, and these parts are *substances.* That which endures through change is a part of a world. Interestingly, then, a world is not itself a substance, but a composite of substances. No substance is for Kant a world. For example, he says, "the so-called world *of the ego,* which is completely constituted by a unique simple substance together with its accidents, is not properly called a world, unless it is called an *imaginary* world."

The *form* of a world is a bond established by the principles of "the *possible influxes* of the substances which constitute a world." Actual influxes are *states* of the world, and "transeunt" forces, the causes of the influxes, presuppose a principle by means of which each independent substance is related to others. In the absence of such a principle, the positing of transeunt forces would be impossible. A world form is thus a set of possibilities maintained throughout changes in the actual states of the world. The identity of the parts is not sufficient for the identity of the whole; the identity of the whole is only insured by a principle of composition. Each "real" world will be given in a form that is constant and invariable, and this form will act as the "perennial principle" of each contingent and transitory form that is a *state* of the world. Kant concludes that the actual world is the world whose changing states are invariant with respect to the forms of space and time, as precisely the forms of receptivity of all singulars of intuition. The compositional identity of World II is space/time identity. It must be noted, however, that the states of possibility are only generated by *given* forces, actual forces empirically ascertained. Kant's insistence that forces can only be given in experience, and cannot be cognized *a priori,* is urged in all stages of the development of his philosophy. This being so, can one really wonder why he had to take seriously the claims of a major European scientist to direct access to the spiritual lives of others?

In 1770 Kant's theory of worlds yields the following structure:

1. Logically considered, the members of a world consist of a set of self-dependent substances.
2. A logical set of substances is not itself a world. A world is a composed or related set of substances, and the relations

binding substances together in a world can only be external relations.

3. The form of such external relatedness is a set of possible interactions (influxes).
4. In any world accessible to human knowers the actual interactions between substances are empirically noted, and are *states* of the world. In the *Dissertation* the distinction between substance and state, between form of world and principle of determining state of a world marks the new departure from the Leibnizian concept of world in *Living Forces*.
5. Whatever transeunt forces we postulate as causes of states of the world must comply with the form of the world.
6. The form of a world is not given independently of the world, but is a condition on what it is possible for us to know in and of that world.
7. The interdependence of self-subsistent substances is guaranteed as a causal nexus supplied by the forms of space and time.
8. These forms do not determine the content of a knowable world, but insure that any forces to be countenanced in the sensible world must be given in empirical intuition.
9. The form of any knowable world is space/time. The principle associated with this form rules out intellectual intuitions and spiritual forces, except as logical possibilities.
10. World I is thus uninstantiated. Its form is that of deductive logic. It exists only as a logical possibility.

How many worlds are there of type II? Only those for which there are forces. Kant had written in the second letter to Mendelssohn that his general problem was one of understanding any case of external impact upon that which is merely receptive. External impact is exerted by the transeunt forces,[24] and these can only be known through sensation organized by the receptor forms of space and time. Kant's World I, as a creature of pure thought, is symbolically cognized (sect. 10); it cannot be represented intuitively -- we have no data to instance it. The matter of our congitions is only given in sensation. World II is thus the only world that can be actual *for us*, actual as a *phenomenal* world of sense-intuited objects organized in space and time, and driven by physical forces empirically ascertained. Kant's important conclusion in the dissertation regarding evidence or data -- the central problem marked in the letter to Mendelssohn is that there are two acceptable forms of intuition. Immediate singulars are given in *sensory intuition,* and in *a priori* or *pure intuition* (a form of direct presentation of particulars later to be rendered by Kant as mathematical *construction*). There is no *intellectual intuition* that would give us immediate access to World I.

The direct bearing of Kant's first attempt at a solution of his "general problem" on the question of the existence of spirits (and hence the knowability of spirits) is not far to seek. The presumed existence of spirits is introduced surreptitiously, by inappropriate -- not intuitively instantiated -- appeals to rational principles, for example, the principle of non-contradiction or identity on which Leibniz had put such great emphasis. In *Inaugural Dissertation* (sect. 28) this is what Kant says:

> But by treating the subjective conditions of judging as objective it is rashly concluded that for that reason it would not be permitted to any intellect at all and so, that *whatever does not involve a contradiction would be therefore possible.* Hence so many vain fabrications of *forces* I know not what invented at pleasure, which freed from the obstacle of inconsistency burst forth in a horde from any spirit which is architectonic or if you prefer it with a proclivity for chimaeras. For since a *force* is nothing else but the *relation* of a substance A to *something else* B (an accident) as of the ground to the grounded the possibility of each force *does not rest upon the identity* of cause and caused or of substance and accident. And so also the impossibility of falsely fabricated forces *does not depend upon contradiction alone.* So it is not permitted to take any *original force* as possible unless *it has been given by experience,* nor can its possibility be conceived *a priori* by any perspicacity of the intellect.

Kant's ultimate appeal to well-ordered judgement under the strictures of DGM occurs in sect. 30:

> ...the ill-considered citation of things supernatural is the couch upon which reclines a lazy intellect. For the same reason, *comparative miracles,* namely influxes of spirits, we carefully keep out of the exposition of phenomena. For since their nature is to us unknown the intellect would be turned aside to its own great detriment, away from the light of experience by which alone it has the opportunity of procuring for itself the laws of judging, towards shadows of species and causes to us unknown.

The only worlds we can know are worlds forced upon us. The principle of non-contradiction (identity) can only establish logical possibility, not real possibility. Leibniz knew this, and sought to have compossibility supplement logical possiiblity. For Kant, the primary criterion of compossibility is the order of actual influxes. We have no experience of spiritual influx, therefore there is nothing to ground a spiritual world except as a logical possibility, a possibility of pure reason. For Kant, the intellectual world (World I) stands only as a kind of logical surrogate for a world of actual spirits. The cognitive status of that world must therefore be radically rethought, a task Kant undertakes in the *Critiques*.

Note the interesting twist. By an application of his special concept of a possible world Kant effectively eliminates the spirit world as a possibly actual one. Possibility can only be established by actuality (perceived forces), and this is why Kant insists that there be some data to underwrite the claims to existence of a spirit world -- this I take it is why he was so fascinated by the alleged clairvoyance of Swedenborg; he wanted to learn if these cases are authentic, because if they are, he would have to give up his moves away from Leibniz, and conclude that two structures of possibilites -- two worlds -- are actualized, the phenomenal and the noumenal: in other words, he would have to accept the metaphysics of Leibniz on the evidence of direct contact between embodied souls and disembodied ones. What he does conclude -- in the absence of evidence of spiritual forces -- is that the two worlds follow from two uses of reason, one theoretical, one practical. Perhaps he fails, but the attempt is a consistent application of DGM, and is entirely faithful to the methodological motivations of Leibniz.

The Corupus Mysticum. In the *Critique of Pure Reason* Kant introduces his special *moral world* (call it World III) as a transformed World I and World II combined, or as a form of World I seen from another point of view. The datum of World III is human freedom -- human freedom is, if you will, the influx force of the moral world, that which gives it its form. If we subtract from this world all actual moral content, it is an intelligible world only (a world of type I). Kant says:

> To this extent, therefore, it is a mere idea, though at the same time a practical idea, which really can have, as it also ought to have, an influence upon the sensible world, to bring that world, so far as may be possiible, into conformity with the idea. The idea of a moral world has, thefore, an objective reality, not as referring to an object of an intelligible intuition (we are quite unable to think any such object), but as referring to the sensible world, viewed, however, as being an object of pure reason in its practical employment, that is, as a *corpus mysticum* of the rational beings in it, so far as the free will of each being is, under moral laws, in complete systematic unity with itself and with the freedom of every other. (A808/B836)[25]

We are now in a good posiiton to see how Kant works out the status of his World III in ways generally faithful to DGM. The interests in that world are typically moral interests; they are subject to the demands of practical, not theoretical, reason. As an idea, the moral world is also an *ideal:* activities in the sensible world are to be adjusted to its demands. We need not stumble over the details of Kant's exegesis of World III in the 2nd *Critique* and the later ethical writings. My point here is just the one that in his attempt to deal with Swedenborg's alleged experiences of the spirit world, and with his spiritual doctrines more generally, Kant was extending his deep commitment to DGM to the fullest measure, even to the point of reaching what Leibniz referred to as the apex of ethics and metaphysics. At one crucial level of his thought, Kant identified the interests of metaphysics with those of ethics.

Two appeals to the arational are made in all of this: an appeal to the datum of human freedom (corresponding to the appeal to sense intuitions in the case of World II); and an appeal to the fundamental and unchallenged rightness of the Double Government Methodology.

The Reasonableness of Kant's Interest in Spiritualism. I now take it that my case has been made. A proper understanding of Kant's interests in the para-normal can be had by close attention to certain details of his intellectual career in the 1760s, understood against further developments of that career in Kant's mature critical period. Given this understanding, no appeal to external psychological and sociological factors is required. Kant's final verdict on the Swedenborg allegations *was* equivocal, but in entirely defensible ways, if we take the arational factors into strict account. There is a World III, but its inhabitants are all living in World II. To take World III to be itself natively inhabited is to take World III to be a World I with intuitive content. For kant, there is no such World I. However provocatively *Träume* may have been styled by Kant, the message of this book, and its intended impact, is systematically and practically related to the contents of the second letter to Mendelssohn, to the fragments referring to Swedenborg and to the lecture notes, to the *Inaugural Dissertation,* and to the later critical works. It is clear, furthermore, that Kant *had to be* concerned with claims to ESP. For if there are spiritual forces, if spirits can be given a habitation, then Kant's entire critical programme collapses in shambles.

Replacing the Broad Explanation. I will try to draw together the threads of this complicated challenge to the strong programme in sociology of knowledge. In seeking such culminated interlacements philosophers often appeal to formalisms. In the case, no exotic notation is required, for we have the Gutting explanatory schemata ready at hand. What needs to be accounted for is Kant's complex attitude toward the question of the existence and knowability of spirits, an attitude seriously coloured by the tone and anonymity of *Traume.*[26] Kant's conlcusion may be stated as follows:

1. The existence of spirits is provably unknowable, although that existence is a *logical* possibility. However, claims to the factual existence of spirits must be taken seriously, because such claims are based on a surreption of actual perception, and this procedure must be exposed, isolated, and the damage done by it corrected. [I take it that this is Kant's full conclusion of the matter. The tone and style of *Träume* appear to call this view into question, or at least give one grounds for thinking that Kant's views on ESP and related paranormal phenomena are equivocal or unsettled.]

A Gutting type II explanation of this conclusion (call it a "Broad Explanation"), offers the following explanans:

1.1 An East Prussian *privat dozent* living in the mid-18th century needed to address himself only to serious and accepted scientific concerns, and needed suitably to disguise any interests in academically marginal or proscribed topics like ESP and the existence of spirits. Failure to comply with such fraternal requirements might well mean non-appointment to a professorship.
1.2 Serious scholars who spend large sums of money for boring books which are not the ones they expected to receive often take a negative view of the sentiments of the authors of such books, and write abusive responses.
1.3 Kant found himself in both sets of circumstances.

I have argued that the Broad Explanation fails to do justice to Kant's attitudes in the *Träume*, and his views on spiritualism in general. The Broad Explanation fails because it is based on an incomplete historical study, and because it ignores Kant's standing commitment (an at least partly arational commitment) to the Double Government Methodology and its implications. I would therefore propose the following explanans as fully capturing what I take to be Kant's conclusion:

1.4 Kant's explanation of DGM led him to expect that explanation of alleged paranormal phenomena would have to be cast in mechanical terms, or that such phenomena, failing of explanation in those terms, would turn out to be products of surreption, meaning that the true status of the phenomena would therefore have to be radically rethought.

1.5 Kant discovered that spiritual phenomena are surreptitious -- that the data of the existence of spirits could not be regarded as reliable. Nevertheless, spiritualism has much popular support, and threatens to subvert orderly morality by encouraging irrational motivations.

From which 1. follows.

Why did Kant take the allegations of Swedenborg's spiritualism seriously in the first place? I have dealt with some details of the answer above. The answer involves recognition of the following: [1] Kant's meta-reason, his acceptance of DGM, already committed him to some form of explanation of "spiritual" phenomena; [2] Kant's persistent empiricism led him to insist on clear evidences (data) of existence; *because* [3] his motivating problem in the 1760s was one of impact (influx) on that which is merely receptive; [4] the preformation of doctrines later to be given full expression in the 1st *Critique* required a principle of demarcation separating authentic cognition from metaphysical pretense (recall that in the letter to Mendelssohn Kant refers to his new "important insights in this discipline [metaphysics] ..., insights that will establish the proper procedure for metaphysics. My notions are not merely general ones but provide a specific criterion."; and [5] even as late as the 1790 letter to Borowski Kant was disturbed by the *Schwärmerei,* the irrational enthusiasm that threatened to defeat religion within the bounds of reason alone, religion sharing the apex with ethics and metaphysics. We thus have reasons and meta-reasons aplenty to account for Kant's quite legitimate interest in the unseemly side of metaphysics.

Assimilating Meta-reasons into Rational Explanation Schemes. A single case like the one I have just investigated in detail cannot, of course, sustain the general case. Was Kant's approach to the problems of spiritualism "rational"? Yes -- and no. It certainly was rational, and progressive, in Laudan's senses of these terms. It tended, within Kant's enlarged research tradition (including importantly Leibniz' DGM), toward solution of a number of problems that would have remained unsolved within the limits of the alternative tradition

(Swedenborg's spiritualism).[27] Still, it must be admitted that some of the features of Kant's approach to these problems were not "internally" rational in the sense that for all of his beliefs Kant had given, or could give, or was prepared to give, a reasoned defense. Some of those beliefs were "basic," or arational. But they go along with the entire programme or tradition, helping to make it serve its larger purpose. Finally, what emerge as constitutive elements in Kant's developing beliefs about the paranormal call for no independent (sociological) explanation, once the intellectual history is written in full detail, as I have endeavored to do. In the best sense of the term I can think of, Kant had *reasons* for taking on Swedenborg, and for developing the view of spirits that he did. None of those reasons were uniquely sociological.

Earlier in this essay I suggested that meta-reasons are utilities, ends in view. In my analysis of the episode in Kant's intellectual life, many such utilities appear. All five of the requirements of Kant's programme outlined above involve utilities. Some are local, like attempted satisfaction of the demands of the problems currently exercising his mind. Some are writ large, like his long-term commitment to satisfying the (religious, ethical) demands of practical reason. On my view of what is required of intellectual history, these utilities become part of the story to be told, *within* the processes of thought of scientists. Finally, I think that Laudan (and James Brown, and others) is largely right in insisting that explanations of scientific belief must conform to some (correct) theory of rationality. If this is *my* methodological meta-reason, so be it.[28] Acceptance of such a meta-reason contributes importantly to awakening from the present *Träumerei* promoted by too much attention to the claims of the strong programme in sociology of knowledge. One would be wrong, however, to suppose that all enthusiasms are pathological. The strong programme is not pseudo-science. In shifting attention away from utilities and meta-reasons to scientifically unfounded generalizations, it presents a methodology that distorts intellectual history.[29] Sociology is by no means redundant in the attempt to understand that in some respects scientific work is surely socially conditioned. But such sociological explanation of some features of the scientific enterprise must, I think, take its place *beside* the explanations of intellectual historians that are based on rationality assumptions or meta-reasons. If science is regarded as a problem-solving activity, then there must be some point at which we can say that problems have, or have not, been solved. I tend to think myself that at that point pragmatic considerations come to play the most important role, and thus my own methodological reconstruction of the progressiveness of science ultimately differs from that of both Laudan and Bloor. I have told some parts of that story elsewhere.[30]

Department of Philosophy
University of Western Ontario

Notes

*The material in this essay dealing with Kant and spiritualism is more fully discussed in Chapter III of my book, *Kant and the Double Government Methodology: Supersensibility and Method in Kant's Philosophy of Science,* Dordrecht: D. Reidel (1984).

1. In *Historical and Philosophical Perspectives of Science,* ed. Roger Stuewer; *Minnesota Studies in the Philosophy of Science,* V (Minneapolis 1970). A quite different, but promising approach to these problems is to be found in James Brown's doctoral dissertation, *Models of Rationality and the History of Science* (University of Western Ontario 1981).
2. Maybe what I am here calling meta-reasons are Gutting's "basic beliefs." I'm not sure about this. I do think there are such creatures as basic beliefs, and that these can be described following identification. If all that the strong programme advocate needs is the assimilation of basic beliefs into my kind of Type I meta-reason explanation, then that programme begins to me to look quite innocuous, because then the scientific questionableness of Type II explanations need never arise, and the absence in sociology of knowledge of any established law-candidate generalizations need cause no embarrassment. I am not suggesting that what I am calling meta-reasons cannot

themselves be argued for. Often they are revealed as suppressed premises in the explanation, but are themselves conclusions of independent reasoning.

3. *Progress and Its Problems* (Berkeley & London 1977), p. 208. I am afraid that Laudan will not be completely satisfied with what I am up to: showing that the arational functions in rational explanations.
4. The exception is the very full and competent study by C.D. Broad, "Immanuel Kant and Psychical Research," *Religion, Philosophy and Psychical Research* (New York 1953), pp. 116-155. I will rely heavily on Broad's clear analysis of the case, but will have to update the information he had available to him.
5. In a postscript to his essay Broad adds to the literature, introducing reference to Kant's *Lectures on Rational Psychology,* in which Kant mentions Swedenborg with approval. The date of the letter to Miss von Knobloch is disputed; I think Broad's dating of 1763 is likely correct.
6. That Kant made regular reference to Swedenborg in lectures on metaphysics is now amply evidenced by published transcripts of notes of Kant's students. See *Metaphysik L1,* Kant's *Schriften,* 28.1, pp. 298-99; *Metaphysik L2, Schriften,* 28.2.1, p. 593; *Metaphysik Herder, Schriften,* 28.1. pp. 113-14 (compare *Nachtrage Herder, Schriften,* 28.1, pp. 857-88); *Fragment einer späteren Rationaltheologie nach Baumbach, Schriften,* 28.2.2, pp. 1324-25. Additional references are in *Strife of the Faculties,* Ak. Ed., Vol 7, p. 46; and in *Anthropology,* Ak. Ed., Vol. 7, p. 191. See also *Reflexionen* 1486 and 5026, the latter containing this fascinating claim: "That on which the plausibility of a metaphysical hypothesis (Swedenborg) rests. On a putative intellectual intuition on analogy with the sensible one". Can it be that Swedenborg is the source of a doctrine (there is intellectual intuition) that Kant will finally put aside decisively in *Critique of Pure Reason?*
7. February 7, 1766, and April 8, 1766. The earlier letter is not extant, but is quoted in *Kants Werke,* ed. Rosenkranz u. Schubert, XI, I (1842). The second letter is translated in Arnulf Zweig, *Kant, Philosophical*

Correspondence 1759-99 (Chicago 1967), pp. 54-57.

8. Kant had purchased, for £7 sterling, the eight large quarto volume set of Swedenborg's *Arcana Coelestia,* a work subjecting every word in the Biblical books of *Genesis* and *Exodus* to a symbolical interpretation. Broad says "It may fairly be described as one of the most boring and absurd productions of any human pen." Kant read the volumes. It is tempting to think that he took the Schwärmerei to have been generated by this kind of reading! Twenty-four years after writing the *Träume* Kant was still worrying about spiritualist ecstacies.
9. It is a long letter, running to nearly 2,000 words. Ludwig Ernst Borowski reprinted the German original of the letter in his *Darstellung des Lebens und Charakters Immanuel Kants* (Konigsberg 1804), pp. 211-25.
10. Not everything Kant says in the letter is reserved. He says of the incident of the Stockholm fire, "The...occurrence appears to me to have the greatest weight of proof, and to place the assertion respecting Swedenborg's extraordinary gift beyond all possibility of doubt."
11. I am referring to the neglected third part of this celebrated treatise, *Enthalt eine Vergleichung zwischen den Einwohnern der Gestirne,* where Kant's thinking extends beyond the mechanistic framework of the rest of his book. He claims for these speculations only that they are not arbitrary, though he also adds that the claims are "not quite indubitable." The publisher of this book declared bankruptcy, and nearly all copies were impounded until 1765 or 1766. There is very little evidence that Kant's close contemporaries knew much about his important astronomical speculations.
12. It is a historical curiosity that Swedenborg himself had introduced a form of the nebular hypothesis 21 years prior to the appearance of Kant's *Allgemeine Naturgeschichte,* in the *Principia* of his three volume *Opera Philosophica et Mineralia* (1734). Beginning in July, 1736, Swedenborg travelled extensively in Europe. In the notes on his journeys, he records that sometime in 1736 he began the study of Wolff's philosophy, had made his

acquaintance, and corresponded with him from time to time. I am not a close student of the Swedish magus, and so I do not know how deeply Swedenborg was influenced by Leibniz, if he was at all. The *Principia* gained for him a European reputation. There is no evidence that Kant had read the work. Kant is not always faithful in naming sources, but in the case of the *Allgemeine Naturgeschichte* he is careful to say that an account in the Hamburg publication, *Freie Urteile* (1751), of the work of Mr. Wright of Durham, had suggested that the fixed stars form a system resembling that of the planets. [The Hamburg account appears in translation in full in *Kant's Cosmogony*, trans. W. Hastie; revised and edited, Willy Ley (New York 1968), pp. 166-79.] I think it fair to assume that if Kant had known Swedenborg's work he would have credited it. Finally, it was not Swedenborg's natural philosophy, but his commitment to the paranormal -- which he took to be quite normal -- that became well-known in northern Europe.

13. I supply some of the details of this summary of Leibniz' empirical methodology in "Leibniz on the Side of the Angels," Chapter II of my *Kant and the Double Government Methodology*.
14. A translation of the relevant section is by Paul & Anne Schrecker, *Monadology and other Philosophical Essays* (Bobbs-Merrill 1965), pp. 78-80.
15. Kant is in a strong sense *the* philosopher of method. I would be prepared to argue that the vital nerve of his lifelong ambition is the quest for a rational accommodation of both Cartesian science and Christian longing for a better life by means of a defensible reformulation of DGM. What I have written about Kant's methodology is summed up in "Kant and the Problem of Scientific Methodology," *Proceedings of the Ottawa Congress on Kant*, ed. Pierre Laberge, Francois Duchesneau, Bryan Morrisey (Ottawa 1976), pp. 222-33. The papers on Kant's teachings on hypotheses cited in the Ottawa paper deal with Kant's attempt methodologically to eliminate a science of the supersensible: "Hypothesis and Explanation in Kant's Philosophy of Science, " *Archiv*

fur Geschichte der Philosophie, 43, 2 (1961); "Kant on Hypotheses in the 'Doctrine of Method' and the *Logik,*" *Archiv für Geschichte der Philosophie,* 44, 2 (1962). These papers are now supplemented and improved by the full treatment of methodology in my *Kant and the Double Government Methodology.*

16. Broad's analytic account of the theoretical parts of the *Traume* is extensive and quite excellent; I cannot improve upon it here.
17. The terms 'subreption' and 'surreption' have a common root, and both have the sense of false, sneaky, deceitful. All of this is likewise true of Kant's German term, *"erschleichen"*. An interesting early meaning of 'surreption' has to do with a sin or a temptation secretly stealing unbidden and unexpected into one's mind. Our words also share a common root (meaning 'to creep') with 'reptile' and 'serpent'. One should bear in mind in all of this that the *Critique of Pure Reason* abounds with discussion of what happens when we seek to extend cognition beyond the limits of experience, when, so to speak, idle and inutile thoughts 'creep in'. I think it likely that Kant took this quasi-religious sense of bad epistemology quite seriously. To mark and dispel surreptitious cognition became for him, as we will see, a kind of methodological crusade.
18. I have mentioned Kant's later interest in the *Schwarmerei,* the irrational enthusiasm for things paranormal that seems to sweep through northern Europe in the second half of the 18th century. Hamann, Kant's sometime friend and proofreader for the 1st *Critique,* writes in a letter to Kriegsrath Scheffner that he regards Swedenborg's spiritual paroxysms as a form of "transcendental epilepsy" producing critical frothing. (Nov. 10, 1784) There seems to have been a readiness on the part of many to regard the *Schwärmerei* as a form of mental illness. For Kant the *Schwärmerei* is a corrupt mental state in at least three senses: it is a sickness, a distorted form of cognition, and an inappropriate ethical and religious enthusiasm. Hamann adds in his letter that the entire theory of evidence of the church is derived

from this transcendental epilepsy. He concludes, "Unfortunately dreams and sicknesses are the best data of the energy of our souls." We will see below that this problem of data is for Kant a central concern, a concern full expressed in an important letter to Moses Mendelssohn.

19. Broad (1953), p. 127.
20. The letter is translated in Arnulf Zweig (1967), pp. 54-7.
21. See the references to my studies of Kant on hypothetical reasoning cited in footnote 15 above.
22. My references to Kant's *On the Form and Principles of the Sensible and Intelligible World* are to the translation of G.B. Kerford & D.E. Walford, Kant, *Selected Pre-critical Writings* (New York 1968).
23. I am not proposing a sociological reason for Kant's philosophical style in this work, apart from the explanatorily neutral observation that when engaged in a ceremony one usually follows the rules, or else is not taken to be engaged in that ceremony.
24. It is important to translate Kant's Latin "transeunt" with the same English word, and not to use "transient". The Insel-Verlag German translation gives "übergehende", which is literal and helpful. The OED provides justification for my suggestion, pointing out that "transeunt" has the force of the opposite of "immanent". The transeunt forces operate beyond themselves, causing the influxes. Kant's point is exactly that such transcendent or transeunt forces can only be posited to account for that which is spatio-temporally ordered. The "actual influxes" are thus sensuously ascertained terminations of the work of transeunt forces, and the "possible influxes" will later become Kant's "objects of possible experience". Although he is careful to limit positing of these forces within space/time formed world II -- thus eliminating the specious postulation of objects of intellectual intuition and an inhabited spirit world -- Kant does not remove the suspicion that he is countenancing the existence of objects that are causally efficacious within experience, and yet which actually transcend experience, until the 1787 2nd edition version of the *Critique of Pure Reason*.

Nevertheless, Kant held throughout his career that the only forces we can know are forces given empirically (in *Living Forces* they are known by their effects); see, for example, A207/B252 in the 1st *Critique*. Kant's treatment of worlds in the Dissertation is part of his attempt in the 1760s to bring forces within the limits of sense experience, and under the constraints of the epistemic grammar of space and time.

25. The translation is that of Norman Kemp Smith (London 1950), pp. 637-38.
26. I think I have sufficiently dealt with the question of anonymity above. If there is a residue to be explained, then *this* is a place for addition of sociological considerations; but such considerations will have no bearing on Kant's beliefs regarding the paranormal. Some relevant features of this adjusted account are the following. In his letter to Mendelssohn of April 8, 1766, Kant says that he wrote Träume to prevent people from pestering him with questions about his "prying inquiry into Swedenborg's 'visions'". I take this to be the *occasion* for writing the book, not the *reason*. I have shown that quite a few of Kant's contemporaries knew that he had authored the book. It is one thing to accept authorship of a work dealing with matters of current debate; it is quite another to want one's name affixed in perpetuity to a hastily written book, one with themes not fully developed, one written in an unscholarly style, one in which one's later best arguments are not put forward. These matters receive further treatment in Kant and the Double Government Methodology.
27. Kant's rejection of the knowability of spirits, his deliberate failure to distinguish pathological maladjustment from paranormal experience, and his rejection of any rational psychological solution to the problems of soul/body connectedness, all pave the way for new departures in both psychology and physiology, not to speak of the liberating effects of his thought on ethical theory and philosophy of religion. Swedenborg's spiritualism generated a church.
28. James Brown, in the work cited in note 1 above, seeks to

show that such meta-reasons can be rendered as reasons. I am not sure that he is right, but that is matter for another paper.

29. I am aware that the case study I have offered leaves some conceptual considerations unresolved. Gutting speaks of "basic beliefs", Bloor and Barnes of "interests", and I have introduced "utilities", "meta-reasons" and "arational factors". My terms are closer in meaning to Gutting's basic beliefs (if I understand him correctly). The interests of the strong programme operate outside cases of scientific reasoning and condition that reasoning as "social forces" condition economic developments and social changes. What I have wanted to show is that meta-reasons or utilities operate within scientific reasoning, and furnish one more mark of the value-ladenness of that reasoning. In the context of attempting to understand internal scientific rationality I am inclined to treat social forces as Kant treated spiritual forces: they lack local presence; they are not transeunt.

30. "Scientific Progress: The Laudan Manifesto," Philosophy of the Social Sciences, 9 (1979); and "Methodology and the Functional Identity of Science and Philosoophy," *Probabilistic Thinking, Thermodynamics, and the Interaction of the History and Philosophy of Science; Proceedings of the 1978 Pisa Conference on the History and Philosophy of Science*, Vol II, ed. J. Hintikka, D. Gruender & E. Agazzi (Dordrecht 1981).

Jerry Gaston

ON THE SOCIOLOGY OF BELIEF, KNOWLEDGE, AND SCIENCE

When one agrees to contribute to a volume such as this*, it seems reasonable to assume that defending one's own ideas is a generally acceptable approach to adopt. Whether or not it is worthwhile to do so in a particular instance depends upon the existence of considerable disagreement over sufficiently interesting or controversial ideas. I find the issues at hand to be both interesting and controversial.

This paper likely will not please any of the contributors to the volume, and few if any, of the ideas will evoke much interest. But, on the evidence shown in the Laudan-Bloor exchange, my primary goal is to avoid any misunderstanding about what I say. There is never a guarantee for that, but rather than produce a long set of arguments, I prefer to present a short set of simple points without excessive bibliographical documentation or footnoted explanations to support them. If this were published in a newspaper, therefore, it would go, almost certainly, on the "opinion" page or in a section labelled "correspondence to the editors." But I take some satisfaction because in the continuing debate about the merits of a sociology of scientific knowledge, documentation of "evidence" has not seemed terribly important for convincing anyone.

I wish to discuss these four points. First, I shall indicate the view, repeated so often as to be almost worn out, about the continuing problem with a sociology of knowledge. Second, I shall inquire about a "sociology of knowledge" interpretation for recent development of the strong program in the sociology of knowledge. Next, I shall present my views about the Laudan-Bloor debate. Finally, I shall give an opinion about the current situation in which some of our colleagues in the socio-historical-philosophical studies of science apparently feel more intellectually threatened than ought to be the case.

J. R. Brown (ed.), Scientific Rationality: The Sociological Turn, 249–263.

Traditional Sociology of Knowledge[1]

I assume that most scholars are aware of the long history of attempts at a sociology of knowledge. And this is certainly not the place to detail again that history. But a brief word is desirable to set the stage for the subsequent points I wish to discuss.

As one would expect, there is little agreement on what the earliest scholars really meant when they tried to provide the outlines for a sociology of knowledge. (And, I am not the least interested in trying to provide *the* interpretation.) What seems to have been a reasonable consensus over the years, however, was that early thinkers about the sociology of knowledge wanted to restrict knowledge to that other than the "natural sciences." From the perspective of some contemporary scholars interested in the sociology of knowledge, the earlier position was a mistake. Why did scholars such as Mannheim think it necessary to omit science from the purview of the sociology of knowledge? That remains unclear, although one could speculate. Apparently realizing the corner he was painting himself into, Mannheim (1936) proposed that the solution to the problem of infinite relativism was to claim that intellectuals, because of their special characteristics pertaining to social structural origins, would be "above" it all and they could be the interpreters.[2] I can find neither convincing evidence nor even agreement from others that his proposed solution comes close to resolving the dilemma. And, in my view, I see no reason to believe that intellectuals, regardless of how they may be defined, are immune to influences stemming from their social positions.[3]

Various hypotheses could be suggested as to why the sociology of scientific knowledge was not attempted during the earlier years following Mannheim's work. Perhaps there was resistance to the idea because it seemed too threatening for social scientists (who may have been unsure of their own scientific foundations) to consider the idea that the content of science was socially constructed. Perhaps the number of sociologists who were interested in other topics did not provide sufficient intellectual resources to deal with the problem.

Perhaps the cultural repertoire did not have sufficient intellectual and methodological strategies to deal with the issue. Or, perhaps it was one of many other possibilities.

Many scholars believe that Mannheim's sociology of knowledge was a failure. Hartung (1970, p. 701) writes, "I can only comment that the relevance of the existential determination of knowledge to truth or truth value is still not clearly understood, if for no other reason than that Mannheim has completely failed to show how the validity of a proposition is even partially dependent upon the social situation of the one who asserted it."[4]

Recently, Bloor (1976, p. 14) challenged these positions and then briefly summarized his position as this:

> Because these arguments have become so taken for granted their formulation has become abbreviated and routine. They can now be given in such condensed versions as the following, provided by Bottomore: "For if all propositions are existentially determined and no proposition is absolutely true, then this proposition itself, if true, is not absolutely true, but is existentially determined."
>
> The premise, that causation implies error on which all these arguments depend have been exposed and rejected. The arguments can therefore be disposed of along with them, Whether a belief is to be judged true or false has nothing to do with whether it has a cause.[5]

Because of the tradition of unsuccessful attempts at a sociology of scientific knowledge, it seems that the early proponents of that specialty for sociology should not be invoked as the forerunners of today's attempts at a sociology of scientific knowledge. Indeed, if I were trying to develop a strong program, I would rather regard their glaring failure to include science in their domain as an example of faulty thinking than imply that they are responsible (except, of course, in an important negative way) for contemporary attempts to develop a sociology of scientific knowledge.

It is not the case, of course, that the sociology of knowledge as applied to non-scientific areas has been a failure.[6] And its successes are perhaps sufficient reason to attempt a sociology of scientific knowledge. But current attmepts to explicate the social construction of scientific knowledge does not require celebrating the views of those earlier scholars who, if a sociology of scientific knowledge is possible, were simply wrong. It is more appropriate to celebrate the fact that they may have been mistaken than to praise them, for what might turn out to be their dismal lack of insight.

Apparently there have been good reasons (whatever that may mean) for the criticisms of scholars until this point, for surely they are not stupid. But just because the sociology of scientific knowledge has not progressed as far as it might have does not mean that it cannot be promoted now, especially because there are sufficient numbers of brilliant scholars who wish to attempt it. My main, and final observation on the "history" of the sociology of knowledge is this: its success of lack of success since being proposed, and the rehashing all those pros and cons by critics will not matter much. What will matter is how the cumulation of convincing studies can change the conception of contemporary and future scholars. I believe that evidence can be a powerful influence for developing a new paradigm.

Is a Sociology of Knowledge Perspective Relevant for the Strong Program?

If the Mannheim version of a sociology of knowledge has not been totally successful (and there is no general consensus on that), and regardless of whether Mannheim *really* wanted to exclude the natural sciences, there is wide agreement[7] that the sociology of knowledge has been shown to be pertinent to other areas of knowledge. Specifically, it is thought to have potential for the social sciences (see note 6).

Moreover, if the strong program has the requirement of "reflexivity," then we should surely be able to expect a reflexive account of the development of the sociology of knowledge as it becomes a sociology of "scientific knowledge". I would agree with Bloor, that such an account would not render the results of inquiry "wrong" but would provide a better understanding of its efficacy and authority.[8]

Additional sociologial facts suggest the potential for such an enquiry. The "strong" program in the sociology of knowledge is primarily the work of a set of scholars in Western Europe and especially in Great Britain. The geographical location of the source ideas has no bearing on the importance or potential for developing the area, but not even the casual observer can deny the lack of random distribution of scholars working from that perspective. (There are several scholars elsewhere, however; so this is clearly not a localized or parochial set of scholars.)

The chronology of publications calling attention to the desirability of developing a sociology of scientific knowledge could be constructed. But presenting such a chronology is scarcely necessary given that most of us have been alive and well and have witnessed the events. We have shared the common experience of noting each new publication and recognizing the developing trend.

Why this phenomenon should be located primarily in Britain is, as usual, open to considerable debate.[9] But regardless of any other lack of success, the sociology of knowledge tradition surely has been successful enough to offer some working hypotheses to explain *why* non-empirical papers on the subject first became popular for British scholars. Similarly, one would want to know why such papers preceded the actual production of empirical results in the program.[10] And lastly, one would want to explore the reasons why in the new sociology of scientific knowledge, there is so much concern for the philosophy of science and vice-versa.[11]

The Laudan-Bloor Exchange

Because I was not present at the original presentation of these exchanges, I have two advantages: on the one hand, I do not know what, in discussion beyond their prepared texts, was said by Laudan and Bloor; on the other hand, I have their published remarks to study and study again (Laudan, 1981; Bloor, 1981).

As both Laudan and Bloor suggest, discussions such as these are conducted using an imprecise language, concepts which mean different things to different people, and terms which have changed meaning over time. (Often expressions have to be prefaced with adjectives derived from scholars, usually dead, who are unavailable to explain themselves, e.g., "Marxian" X,

"Kantian" Y, or "Hegelian" Z.) Thus, the limitations of language inhibit understanding and agreement. At the same time, one can argue about the accuracy of a position primarily in terms of evidence, the preponderance of which should be pertinent for one or another position; however, if the participants to an exchange cannot agree on what is evidence in the first place, there seems to be little hope of persuasion.

Multiple readings of the Laudan-Bloor exchange produced no clear decision for me about who is (more) right. I can see what may be minor points "won"; but about those I am certain the participants would not agree. Instead of trying to unravel the arguments for the reader and present my "unique" perspective, I shall mention briefly only a few observations.

First, both Laudan and Bloor appear to be of the view that this is an all-or-nothing affair. Bloor claims that the strong program for the sociology of knowledge must meet four criteria (theses) in order to represent a robust scientific approach to the problem of human belief. But Laudan (1981, pp. 180-81) is not ready to accept the claim that

> ...what we call 'the sciences' [presumably including sociology] have any special set of methodological principles or epistemic credentials that clearly sets them off from other supposedly 'non-scientific' forms of cognition. What I am confident about is the claim that no one, philosopher or sociologist, has yet set out any acceptable account of what cognitive or methodological features demarcate the sciences from the non-sciences.[12]

Laudan (1981, p. 181) credits Bloor in that "...Bloor believes not only that there is a clear demarcation to be found [between the scientific and non-scientific], but that he is already in possession of it. He must believe that, for otherwise, he would have no ground whatsoever for claiming his approach to be distinctly scientific." But very soon after, Laudan (1981, p. 181) asks: "Well, what is scientific about theses (1) to (4)?" Then he proceeds to show, without explicitly outlining his criteria, but certainly having some, that theses (1), (2), and (4) are not scientific. Laudan (1981, p. 182) summarizes: "Three of the four theses of the strong program are thus relatively uncontroversial, but their scientific status has yet to be established." And, if that would not require some notion of what is science and what is non-science, the reasoning certainly escapes me. When we get to thesis (3), Laudan gets very concerned that, according to the strong program, the "causes" of any beliefs result from the same causal mechanisms. Laudan (1981, p. 179) indicates that according to thesis (3) the sociology of knowledge would look like this: "it would be impartial with respect to truth and falsity, rationality or irrationality, success or failure. Both sides of these dichotomies," he says, "will require explanation...[.]"

Laudan proceeds to argue that there is no necessary reason why both sides of each dichotomy must programmatically have the same causal mechanisms. He permits the empirical possibility that they might, but he does not see why this should be assumed *a priori*. On this point, I am with Bloor. I would argue that all beliefs (scientific and otherwise) are the result of the same causal mechanism. But I would not base this on a sociology of knowledge perspective or argue this on philosophical grounds.

Rather, I would argue that the answer is in the social-psychological process that we simply call learning. (Many other disciplines, of course, have much to say about learning: the potential for, the speed of, the amount of, the accuracy of, and so forth). We do not know much when we come into this world; and everything we "know" at any particular time, we have learned. To the extent that we learn all content, regardless of what it is, then we must invoke theories of learning to explain how anyone comes to believe anything -- and that includes science. We learn a lot of things, however, that we later believe are not true, accurate, appropriate, good for us, etc. We learn in specially structured environments, we learn to accept what others tell us, and we also learn "out behind the barn." All the many different ways we learn are indeed the ways we learn. *What* we learn, as I would assume everyone agrees, is different from *how* we learn. And it is at this point that I am much more ready to agree (presumably with Bloor) that coming to "believe" something about science is highly dependent on the specific learning environment present at the time.

The structure of learning, to me, then, seems to hold the key to the empirical problem of whether or not scientific beliefs are causally determined by the same sociological "reasons" that determine other beliefs.

I am well aware that suggesting another level of inquiry (social-psychological) is not going to contribute to the Laudan-Bloor exchange. But that leads me to my third and final point about their debate: I cannot understand why Bloor wishes to convince philosophers about the "strenth" of the strong program, or why he attacked philosophers instead of primarily sociologists. Who, as a practicing sociologist, should really be concerned whether philosophers are convinced? And, equally important, why should philosophers care what sociologists think? Neither group is going to convince the other. Sociologists do not agree with each other. And these intradisciplinary debates presumably are conducted within a reasonably common set of assumptions. Though the assumptions are not identical, they are more likely to be understandable to members of the same discipline than to outsiders. And what if philosophers have the strongest arguments and evidence? Is that going to change sociologists' practice? And if "truth" is not to be found (because it is irrelevant), about what are we debating?

Now I am not -- I repeat not -- suggesting that philosophers and sociologists should avoid each other. The sparks created are too much fun. What I am suggesting is that we admit that little, if anything, is going to change. It is like an apple and an orange debating the question of what is the appropriate color for a piece of fruit!

Perception of Intellectual Threat in the Sociology of Scientific Knowledge

In the second section of this essay, I briefly inquired about the pertinence of the sociology of knowledge to explaining the "new" sociology of scientific knowledge (strong program). And, I pointed out the lack of random distribution of interest in the sociology of scientific knowledge.

Having spent about the last fifteen years as a presumed member of the sociology of science community, I have observed some activities which at times have been rather unnecessary and therefore disturbing. Some of these activities have occurred in printed publication; other times the action has been through formal oral publication; and not a little bit has transpired as interpersonal verbal assaults. I am referring, of course, to the perennial, and almost continuous, "intellectual" conflict that has been obvious in the social studies of science.

Let me try to avoid being misunderstood: in no way am I suggesting that intellectual disagreement should be repressed; in no way would I propose that everyone ought to agree; and in no case would I suggest that organized skepticism should not be prevalent. What I do think is desirable is a bit of fairness and accuracy. (Some would add, civility).

The main goal of some scholars in the social studies of science apparently has been oriented more toward eliminating diverse views than toward nurturing them. This shows in the tone and character with which various debates have been conducted. Wholesale labeling of "Mertonian" sociology (or sociologists) of science as natural enemies of a sociology of knowledge has become commonplace.

At one point (and prior to the more recent developments in the new sociology of knowledge), the appearance of Thomas Kuhn's work offered a view presumably opposite to the "standard" "Mertonian" view. This seemed to produce an amount of glee that ordinarily comes from more earthy pleasures. But as time has passed, considerable evidence suggests that the Kuhnian view was not so antagonistic as was originally thought.[13] And this is especially important because neither Kuhn nor Merton -- on my reading -- ever assumed (much less claimed) to have the total and final word about "science."[14]

So, to a large extent Kuhn enhanced our sociological perspective by reminding us how sociologically pregnant some processes are in science; but the Kuhnian perspective failed to eliminate or eradicate the "Mertonian model."[15] Now, anyone who has studied carefully the Mertonian model and the literature that grew from it knows -- but many apparently will not admit -- that Mertonian sociology of science never claimed the far-reaching implications others attributed to it. It did not attempt to be all things to all people. And, throughout all of Merton's own writings (with which I am rather familiar)[16] there is not a single place where I can find that he suggests or explicitly claims the comprehensiveness that others have tried to attribute to him. Indeed, Merton can always be found proposing additional approaches that might shed light on an issue.[17] That practice could hardly support such an extreme conclusion that he presumed his work in the sociology of science to be comprehensive.[18]

Even more to the point, long before most of us ever started more serious study in the university context, Merton (1957, p. 508) published these words in concluding a review of Mannheim's ideas:

> The critical tone of the foregoing discussion should not be misleading. Mannheim has sketched the broad contours of the sociology of knowledge with remarkable skill and insight. Shorn of their epistemological impediments, with their concepts modified by the lessons of further empirical inquiry and with occasional logical inconsistencies eliminated, Mannheim's procedures and substantive findings clarify relations between knowledge and social structure which have hitherto remained obscure. Fortunately, Mannheim recognizes that his work is by no means definitive -- a term which strikes a harsh discord when applied to any work of science -- and we may await considerable enlightenment from further explorations of the territory in which he pioneered.

But in spite of the evidence to the contrary, a constant flow of papers has tried to misrepresent his position so that it could be demolished. One can only say, most of these are not innocent misinterpretations, not errors of perception: they are clearly wrong and in some instances blatantly wrong. Such behavior, of course, raises a question about the motivation of some talented scholars.

Now, you may ask, why am I so "defensive." I have not hesitated aboved to indicate that fairness is desirable, and the least that we should expect. Specifically, not only are there ample examples of the lack of fairness, but once a critical attribution has been directed toward "Mertonianism," one then sees that others -- including the likes of me -- are included in a broad generalizing sweep.

As a consequence of these distortions, which one can only assume to be deliberate (because those who distort are brilliant and know better), there are three items which annoy me. *First,* there is the misrepresentation which is simply not necessary for the argument. (About the "interests" in this kind of sport, one can only speculate). *Second,* there is the view that insofar as many of us thus far have worked on the institutional approach, we are opposed automatically to other views, and are hostile to holders of "non-Mertonian" views. This alleged natural opposition by "Mertonians" is supposed to make the other side's position even stronger. (The thinking seems to go like this: Here is my argument, Y; you don't approve of X; X is certainly against Y; so I know you can more easily accept Y.) *Third,* there is the allegation that, having yet to address the *real* issue -- the sociology of scientific knowledge -- those in the "Mertonian" perspective must be unwilling to admit efforts in that area as legitimate concerns.

The reality is that most of us had not got around yet to developing an acceptable methodology for such studies, whereas some scholars (generally outside of North America) for several began their research program earlier. In the process, their work apparently required a posture which defined the only real sociology of science as the sociology of *(scientific)* knowledge. Proponents and active scholars in the strong program, in attempting to demolish alternative views, are not reticent to claim or assert that *the* sociology of science must be a sociology of scientific *knowledge*.

It is difficult to understand why some of those who prefer the strong program would try to argue for a monolithic perspective. In part, I suspect this may be a response to the totally inappropriate opinion held by some, that in the earlier years of the sociology of science (say, from about 1955), the functional perspective generally *exercised* some kind of monopoly. The demonstrable fact is, however, that at no time were scholars in the perspective proposing or implying that it was the only interest for the sociology of science.

In contrast, I believe that any strategy (disciplinary, theoretical, or methodological) which tries to reveal more about science -- historically or comparatively -- is a worthwhile effort. If some scholars want to approach the study of science from the perspective of the strong program, they should do so. If others wish to pursue research using other strategies, they should do so. It is important that we see how successful the strong program can be. And even if it is more successful than even its strongest proponent expects, much about the phenomenon called "Science" will still require investigation.

Let's get on with it.

Department of Sociology
Texas A & M University

Notes

*I was unable (because of prior commitments) to participate in the conference which culminated in this volume. So the present essay was written without benefit of the discussion which occurred at that conference.

1. This is not intended as a summary, an effort that would not be possible in the available space. Several summaries are available, but for a well-developed summary, consult Mulkay (1979) and for a variety of views, see Curtis and Petras (1970).
2. This is an all-too-concise statement of what Mannheim (1936) said.
3. For a rather witty statement about this, see Popper (1970, pp. 652-53).
4. Several criticisms, devastating at least on the surface, may be found in Hartung (1970), de Gre (1970), and Popper (1970). Whether they are more than that will depend on one's views about the sociology of knowledge.
5. I am unusually hesitant to quote anyone briefly in this type of essay because of the obvious possibility of inadvertently quoting something out of context. But I risk that anyway.
6. At least, one could argue that there was some success as shown by Waltung (1970) and Gaston and Sherohman (1974).
7. Mulkay (1979, pp. 15-16) argues that Mannheim (1936) almost included the sciences in his array of cases for the sociology of knowledge.
8. This is elaborated considerably in Bloor (1976, p. 73).

9. Ben-David (1978) provides a provisional set of hypotheses for this phenomenon.
10. One could ask further: "What theoretical perspective might explain the apparent lack of negative findings in the empirical research thus far?" But, exploration of that deserves fuller treatment another time.
11. The Laudan-Bloor exchange is only one incident, of course. Mulkay (1979, p. 61) writes about the new philosophy of science and how it has produced an opportunity to rethink prior assumptions about a sociology of knowledge.
12. I hear a lot about what is science and what is non-science. Indeed, Laudan's (1981) title is "The Psuedo-Science of Science?" which asks: is the strong program pseudo-science? If the answer is "yes," then we have a non-science or at least something that does not measure up to science. If the answer is "no," then we have something that meausres up to our notion of science. What puzzles me is why, over the last three centuries, can we still not agree on the features we will use to demarcate science from non-science.
13. For some further ideas on this subject, see Gaston (1978, 1980).
14. For an appreciative treatment of Kuhn, see Merton (1977).
15. The perspective failed, not because Kuhn tried; it failed because others' efforts did not succeed in their use of it.
16. Others who may wish to become familiar with the breadth of Merton's work might consult Mary Miles' (1975) extensive bibliography.
17. The extensive reader of Merton will recognize this evidence more readily when I mention the exhaustive (and sometimes exhausting) footnoes in most of his works. The example *par excellence* is Merton's (1965) *On the Shoulder of Giants.*
18. Because I have never discussed this topic with him, I rely on printed texts.

References

Bloor, D.: 1976, *Knowledge and Social Imagery,* Routledge and Kegan Paul, London.

Bloor, D.: 1981, 'The Strengths of the Strong Programme', *Philosophy of the Social Sciences* 11, 199-213. (Reproduced in this volume.)

Ben-David, J.: 1978, 'Emergence of National Traditions in the Sociology of Science', in J. Gaston (ed.), *Sociology of Science,* Jossey-Bass, San Francisco, pp. 197-218.

Curtis, J.E. and J.W. Petras: 1970, *The Sociology of Knowledge,* Praeger Publishers, New York.

Gaston, J.C.: 1978, *The Reward System in British and American Science,* Wiley-Interscience, New York.

Gaston, J.C.: 1980, 'Sociology of Science and Technology', in Paul T. Durbin (ed.), *A Guide to the Culture of Science, Technology, and Medicine,* Free Press, New York, pp. 465-526.

Gaston, J.C., and James Sherohman: 1974, 'Social Origins of Researchers on Black Americans', *The American Sociologist* 9, 75-82.

de Gre, G.: 1970, 'The Sociology of Knowledge and the Problem of Truth', in J.E. Curtis and J.W. Petras (eds.), *The Sociology of Knowledge,* Praeger Publishers, New York, pp. 661-667.

Hartung, F.: 1970, 'Problems of the Sociology of Knowledge', in J.E. Curtis and J.W. Petras (eds.), *The Sociology of Knowledge,* Praeger Publishers, New York, 686-705.

Laudan, L.: 1981, 'The Pseudo-Science of Science?' *Philosophy of the Social Sciences,* 11, 173-198. (Reproduced in this volume.)

Mannheim, K.: 1936, *Ideology and Utopia,* Harcourt, New York (trans. by Louis Wirth and Edward Shils).

Merton, R.K.: 1957, *Social Theory and Social Structure,* Free Press, New York.

Merton, R.K.: 1965, *On the Shoulder of Giants: A Shadean Postscript,* Free Press, New York.

Merton, R.K.: 1977, 'The Sociology of Science: An Episodic Memoir', in R.K. Merton and J. Gaston (eds.), *The Sociology of Science in Europe,* South Illinois University Press, Carbondale, pp. 3-141.

Miles, M.W.: 1975, 'The Writings of Robert K. Merton', in L.A. Coser (ed.), *The Idea of Social Structure*, Harcourt Brace Javanovich, New York, pp. 497-522.

Mulkay, M.: 1979, *Science and the Sociology of Knowledge*, George Allen and Unwin, London.

Popper, K.: 1970, 'The Sociology of Knowledge', in J.E. Curtis and J.W. Petras (eds.), *The Sociology of Knowledge*, Praeger Publishers, New York, pp. 649-660.

Walton, J.: 1970, 'Discipline, Method, and Community Power', in J.E. Curtis and J.W. Petras (eds.), *The Sociology of Knowledge*, Praeger Publishers, New York, pp. 637-646.

John M. Nicholas

SCIENTIFIC AND OTHER INTERESTS

> Consider the members of two tribes ... Faced with a choice between the beliefs of his own tribe and those of the other, each individual would typically prefer those of his own culture. -- Barry Barnes and David Bloor, 1982.

Strict Relativism versus Contextual Diversity of Opinion

There can be little doubt that relativism is for many philosophers as a red rag to a bull. The introduction of social variables into the characterization of the essential nature of science has been regarded by them with hostility. In some cases, this hostility has been warranted; in other cases not. Part of the problem has been equivocation about the nature of the issue, and part of that equivocation has derived from confusion and vagueness about the key terms of the debate, like "relativism".

The new wave [1] is at pains to emphasize that they embrace relativism wholeheartedly. Whatever intrinsic merits it has for them, their relativism is part of an argument for an essential plank in their platform. This is their claim that all explanations of scientific growth are on a par, independent of considerations of truth and rationality, and that there are no meaningful issues of rationality concerning science. Before coming to the application of their relativism to the furtherance of some of their other doctrines, I wish to attempt to clarify what is at stake for some philosophers in this issue of relativism, and to repudiate the arguments of Barnes and Bloor in particular for what they call relativism.

In the new wave literature it is hard to find a serious attempt to give an exact account of what is meant by "relativism," [2] although it is commonplace to see expressions of wholehearted enthusiasm for the view. Before discussing one of

J. R. Brown (ed.), Scientific Rationality: The Sociological Turn, 265–294.

the few explicit statements of the position, I will begin by sketching what relativism means for many philosophers It will be convenient initially to focus on how relativism is conceived in ethical issues, using that as a ready guide to its epistemological counterparts in what follows.

Ethical relativism is typically understood to denote three families of theory, which differ in crucially important respects from each other. These differences are decisive when translated into their epistemological counterparts which are the crucial ones for this paper. First there is *descriptive relativism*[3] which asserts that individuals or groups hold different ethical values or principles in such a way that disagreements in their ethical judgments will not be resolved by the parties coming to a consensus on the nature of the situations which they are attempting to evaluate. Strictly speaking this means they will continue to hold divergent opinions even when they agree about all matters of relevant fact in their description of the issue. Such divergence may be ascribed to differences of cultural background and training. The position is appropriately termed "descriptive" since it records only the fact that individuals or groups do differ, and it attempts no normative or prescriptive account of which party, if either, is right in its judgments.

Meta-ethical relativism is a normative theory, and not merely a descriptive one, for it asserts that there need not be a single correct moral appraisal of a given issue. That is to say, there may be divergent ethical opinions which are all of them correct in some sense. It is important to note that to embrace this position does not commit one to any particular ethical theory, but it may well be inconsistent with some such theories. This theory is normative in that at least it takes a position with respect to constraints on substantive ethical theories.

A substantive ethical theory which is committed to the view that what a person judges to be right *is* right, or to the view that what is morally right is to be specified for each individual, so that ethical principles make essential reference to the individual's *name* rather than any other attribute, is a member of the family of *normative relativisms*. These positions seem to entail that there would be no point in disputing the morality of an act with a person who unequivocally judges it to be right.

I take there to be a wide philosophical consensus that the discovery of descriptive relativism is logically quite independent of meta-ethical relativism, and of normative relativism. Where one is found, that of itself offers no guarantee that any of the others will be found. In particular, where descriptive relativism is found no prescriptive relativism need follow.

The case is the same in the epistemological counterparts to the above which we may call, respectively, *descriptive relativism, meta-theoretic relativism,* and *normative relativism.* For these counterparts, the same logical independence is assumed. For diversity of opinion (now about matters of empirical fact) does not necessarily reveal any prescriptive relativism.

Having identified, albeit in rough terms, the basic variety of relativisms which concern philosophers, let me turn now to what is claimed by the new wave of relativists. Barnes and Bloor identify as the "starting point" of relativism three doctrines:

1) There is diversity of beliefs in the sense that contrary beliefs are held by individuals and groups.
2) Which of the contrary beliefs on a topic are held in a given context "depends on, or is relative to" the circumstances of the parties.
3) All beliefs "are on a par with one another" as far as their causes go[4] (Symmetry Condition).

Thus, relativism asserts the context dependency of divergent beliefs, and adds a rider that causal explanations of those beliefs have the same generic causes.

The question of the Symmetry Condition of Bloor's Strong Programme, has been discussed elsewhere decisively,[5] and I shall not pursue it further here. Let me turn instead to ask how the first two conditions of this relativism relate to the philosophical categories invoked above, and then ask whether they really are bones which should stick in philosophers' throats. My conviction is that they are philosophically quite palatable.

The most obvious fact is that (1) and (2) are, if they were to constitute relativisms at all, descriptive relativisms. For they assert merely the contingent facts that there is diversity of opinion and that where there is diversity it is correlated empirically with the deliberating agents' differing circumstances. As I indicated above, that there is descriptive relativism shows nothing about normative relativism, and since, in the main, what threatens philosophers is not descriptive but normative or meta-theoretic relativism, this cannot be what is at stake in the conflict between philosophers and the new wave. Whether there are such things as normative questions in epistemology is a matter of dispute, and I will examine that issue below.

But we may have conceded too much. What is described by Barnes and Bloor fails even to meet the basic requirements of descriptive relativism. For a strict relativism should, I take it, disallow context dependency. The parties to the dispute should be able to agree about the relevant characterization of the issue. Where context and background knowledge creep in there is the possibility that the norms which are being followed are norms which require the specification of contextual variables before they have implications for the case at hand. For example, the kind of relativism embraced by Barnes and Bloor can be found where individuals share the same traditional deductive logic, but differ in their contexts of belief. Two scientists may both agree on *modus ponens* and yet disagree on the truth value of a proposition, B. One argues that since A, and A implies B are the case, B is the case. The second agrees about the principle of inference but denies that the second premise is true. As a consequence, B is not the case.[6] Here, however, we do not have a case of *strict* descriptive relativism, for the disputants do not agree over what the epistemologically relevant facts are.

Modern philosophy of science has multiplied the possibilities of the alleged relativism espoused by Barnes and Bloor, which from now on I shall distinguish from strict relativism by the titles "context dependent diversity of opinion", or, more simply, "contextual diversity". In the inductive logics which have emerged in the last thirty years, much latitude has been left for contextual diversity. It has been increasingly recognized that matters of of fact and judgments of rationality do not unequivocally fix epistemic judgment. And it is noteworthy that the motivation has not been the recognition of Kuhnian claims that methodological directives fail to specify unique conclusions to enquiries, but difficulties and insights internal to the programmes.[7]

Within Carnapian and Hintikkan inductive logics are to be found epistemological and ontological prejudices which must be embraced prior to their application. Thus decisions are required concerning the relationship of probability and complexity of the world in order to fix certain of the free or exogenous variables which characterise the families of inductive logic, and which must be settled before any unique values of degree of confirmation can be assigned. Some attitudes to ontological questions must be decided upon at the outset, and individuals who differ in such attitudes -- which cannot be rationalized by the system of inductive logic itself -- may well differ in their respective evaluations of theories in the light of the data. More on the epistemological front, the use of Carnap's lambda families prerequires a decision concerning the relative significance for singular predictive inferences of *a priori* characteristics of the language in which the logic is to be formulated, and *a posteriori* information derived from sampling data. The lambda parameter is effectively an index of the agent's inclination to learn from experience.

Indices of caution and boldness, expressing the deliberating agent's attitude to making (epistemic) errors, abound in inductive logics and decision theories.[8] These, in a general sense, are counterparts to the choice of significance level within conventional hypothesis testing, and to the trade-offs between statistical type I and type II errors.

In subjectivist Bayesian inference, the influence of the prior probability can be considerable, and there are few, if any, rationality constraints on acceptable priors. Even the convergence theorems which indicate the conditions under which divergent priors are replaced by posterior probabilities tending to consensus and true values, require conditions of randomness and independence in sampling data which are not obviously met with in scientific contexts of considerable methodological importance.[9] Even if there is convergence with the accumulation of data, the rate at which it occurs may leave considerable latitude for contextual diversity or inhomogeneity of community judgment. Such inhomogeneity would be the consequence of not irrationally held prejudices interacting in different ways with the assimilation of the very same data.

The same general comments can be made about theories of scientific rationaltiy which view science as a question-answering activity, such as Levi's "local induction" account. For in that approach, decisions have to be made not only about indices of caution with their concomitant attitudes to risk of error, but also about the space of relevant answers which constitutes the so-called *ultimate partition*.[10]

Carnap's most mature inductive logic has some remarkable facets, and should be instructive for those who think that philosophers uniformly shun the possiblity of contextual diversity of opinion. In his so-called Basic System, Carnap clearly identified the sensitivity of the assignments of degrees of confirmation to particular theoretical commitments.[11] The application of the system requires decisions about the structure of the space of attributes which are relevant to the problem at hand. Carnap allowed that if one subscribed to the electromagnetic wave theory of light, then the inductive logic would incorporate this commitment into the similarity relations within the "attribute space" of colours, in a way which would be different from that emerging from a purely phenomenal understanding of colours. This possibility invites an interpretation which would be called Kuhnian by some because we can envisage two "Carnapian" scientists, for historical reasons one a wave theorist and the other a corpuscularian, both using the same generic inductive logic but assigning different degrees of confirmation to selected propositions. Their prior theoretical orientations concerning light lead them to structure the relevant attribute spaces in different ways. The Basic System nicely encapsulates the way in which theoretical commitments are inextricably bound up with methodology. Further, it readily permits contextual dependent diversity of opinion, even though it does not permit strict relativism.

It is worth noting that although the first example above (where diversity arose despite the sharing of *modus ponens* in common) represented the initial divergences in terms of simple differences of accepted beliefs which then were applied as premises in conventional arguments, not all these illustrations have that character. For, even in the case of the Basic System, though divergent theoretical commitments made for diversity of methodological judgments, they did not do so as premises in an argument. They entered in the parametric characterization of similarity in relations. This is even clearer in the case of indices of attitude to risk, and rates of *a posteriori* learning. What seems to emerge from the range of inductive logics available in current philosophy of science is that diversity might have its roots in contextual differences of a type not commonly thought to be epistemologically important previously. Indeed, I suspect that traditionally conceived methodology based firmly on the notion of *logical inference* (albeit inductive) simply could not have permitted these sources of variability. Below, I shall argue that what is at stake in the clash between modern philosophy of science and the new wave is not so much relativism or contextual diversity as it is the particular conceptions of induction thought to be operating in the sciences. In part this is because the interests model which is of such importance to the new wave is highly congenial from the viewpoint of many philosophers of science. For the present, however, it should be noted that what is relativism for new wave sociologists is not relativism in the philosophical sense. The context dependency of the former disqualifies it -- even at the descriptive level. What the sociologists have identified is contextual diversity, and modern philosophy of science, if we understand its theories of induction as at least suggestive of potentially descriptive theories of deliberation, leaves considerable room for contextual diversity of opinion. It does not, as far as I can see, embrace strict relativism.

The Devaluation of Normative Issues

One of the issues which, perhaps predictably, has struck a nerve for philosophers is the attack on the significance of normative questions, the philosopher's staple. It is an attack which makes philosophers even more uncomfortable than squabbles about the aetiology of scientific growth, for it is a direct assault on the notion of rationality, without the mediation of complex views about the relation of descriptive history to prescriptive methodology. The slogan "validity without credibility is nothing" has now come to embody a severe critique of the significance, if not existence, of normative questions and normative answers.

It seems that Barnes and Bloor fear that if some substance can be given to the normative notion of rationality then their symmetry principle, that judgments of rationality or truth should not be permitted to prejudice the character of causal explanations brought to bear on an historical episode, will be undermined. The argument is sketched as follows:

> For the relativist there is no sense attached to the idea that some standards of beliefs are *really* rational as distinct from merely locally accepted as such. Because he thinks that there are no context-free or super-cultural norms of rationality he does not see rationally and irrationally held beliefs as making up two distinct and qualitatively different classes of thing. They do not fall into two different natural kinds which make different sorts of appeal to the human mind, or stand in a different relationship to reality, or depend for their credibility on different patterns of social organization. Hence the relativist conclusion that they are to be explained the same way.[12]

As I indicated in the previous section, most philosophers feel that there is complete logical independence between the existence of true normative judgments and whether, as a matter of fact, any one has lived up to them, for what ever cause or reason. Barnes and Bloor recognize this. At this point, they say, in order "to try to contain and limit the significance of the sociology of knowledge", philosophers declare that the sociology of knowledge is concerned with the causes of historic belief [credibility], while philosophers are concerned with what really is rational in the prescriptive sense [validity]. Barnes and Bloor note that philosophers distinguish "an account of sufficiently good reasons" from "an account of the psychological, physiological or sociological causes of inclinations to utter words expressing this belief when appropriately stimulated". They continue:

> But having separated these two issues this critic proceeds to shunt the sociologist and psychologist into the sidings where they can be forgotten. The rationalist is now free to operate in the realm of reason and make out its

> function and workings to be whatever he wishes. This is why we are told so emphatically that the sociologist of knowledge "must be concerned with causes of belief *rather than* with whatever evidencing reasons there may be for cherishing them..." Unfortunately for the rationalist the freedom which this convenient division of labour would give him cannot be granted; the distinctions upon which it is based will not stand examination.[13]

What Barnes and Bloor offer as "The reason" for the failure of this distinction is "that it would be difficult to find a commodity more contingent and socially variable than the philosopher's (Antony Flew's, in their discussion) 'evidencing reasons'. What counts as an 'evidencing reason' for a belief in one context will be seen as evidence for quite a different conclusion in another context. So, because there exists context dependent diversity of opinion, no distinction can be made between validity and credibility.

Barnes and Bloor then proceed to a conclusion which, it seems to me, they had already been granted, for they propose that if the philosopher were to insist on the distinction between validity and credibility "he would simply leave the field of discourse altogether" and "The sociologist of knowledge ... would win by default: his theory would meet with no opposition".[14] Although they do not intend as much, Barnes and Bloor have inadvertently depicted the scene exactly as the philosopher would have it. For the field is theirs, at least they may carve it up as best they can with the competitors, psychologists, economists, phsiologists, historians, and so on. This was expressly considered by Barnes and Bloor when they quoted Flew's concession, noted above. It is extraordinary that they should see themselves being shunted ... "into the sidings"[15] when their declared domain of interest is conceded to them. For since the beginning, for example, the Strong Programme has been "concerned with the conditions which bring about beliefs or states of knowledge"[16] and "impartial with respect to truth and falsity, rationality and irrationality, success or failure".[17]

What is left for philosophers? Are there really any significant normative questions? The answer is a categorical *yes*. For although there is no clear consensus among philosophers concerning what exactly normative questions are in essence, and what their fine-grained epistemology is, it is at least apparent that there are questions and answers in which they have a legitimate, perhaps even a defining professional interest, and which are quite distinct from the causal explanation of why individuals and groups have the beliefs they do.

An anthropologist contemplates the relationships between beliefs prevalent in pre-literate society and certain social functions or the satisfaction of collective social interests. We approach the anthropologist and ask why he believes one theory rather than another, given his background knowledge and assumptions, together with whatever data he has. He may reply at least two ways. He may reply that the hypothesis that he has formed is the best of those he can think of as far as accounting for the statistical dispersion of sampling results he has put together from his empirical studies, or that a suitable null hypothesis is to be rejected at a 5% level given these data. His response takes the form of the recounting of some basic level of deliberation, with the emergent choice of some best or optimal alternative. Alternatively, he might reply that regardless of the relationship of data, theory and so on, the public espousal of the theory in question will draw much attention in his own society and facilitate his advancement in an intended political career.

Yet again, he may simply say that certain generalities (laws) hold concerning the intellectual classes in the society of which he is a member and he computes that given the initial and boundary conditions which he satisfies he will act as he has. Hence the explanation of his behaviour.

In the first, and appearances not withstanding, second cases, the discussion is justificatory. In the third, it is explanatory. In the first and second cases, question and answer centre on the conformity of the reasoning to certain assumed norms. If the interest centres on the data, the debate is an anthropological debate. If the interest centres on the norms, then the debate is methodological, especially when the question of how good are the norms is addressed. Even though debate about the merits of the norms must appeal in turn to a presupposed base of norms, that debate is not an enquiry into whatever causes there are which bring about compliance to them; it is not an explanatory debate. The examination of the norms presupposed, and the enquiry as to whether they are the correct or best norms is the methodological or philosophical enquiry. What is important about such an enquiry is that it is properly taken to have a different kind of answer from the demand for a causal explanation. It is noteworthy, too, that, while it is undoubtedly the case that each methodological answer is paralleled by a causal answer, each causal answer is paralleled by a methodological issue. For the explanatory force of the third case above, for example, relies upon its conformity to a norm of explanation. It will not be of any avail to substitute an explanatory account for a normative one, whether in the attempted justification of norms or in the invoking of a causal explanation's implicit normative presuppositions; for retreat to a causal level will invoke still further normative presuppositions.

The Underdetermination of Rational Choice

Apart from the exaggerated threat of strict relativism, and the affront of the denial of the existence of philosophy in the disavowal of normative matters, philosophical toes are bruised under the sociologists' ponderous tread when the latter argue for the underdetermination of rational choice. Philosophers have aspired, along with scientists and others who have an interest in finding out the best way to do and to evaluate science, to find canons of inductive inference and decision which do more than simply record the habits of thought of the historical figures who have contributed to science. For the new wave sociologists repeatedly argue that principles of rationality are fundamentally equivocal, and that the gap of equivocation can and must be bridged by appeal to social variables.[18] In the next section, I shall make some concessions to the sociologists' point of view when it is interpreted in a very precise sense. The grounds for this concession emerge from considerations internal to inductive logic and methodology. In this section, I will argue that as long as they maintain that choices in science are propelled by the pursuit of social interests then some model of rational choice must be exempt from the equivocation argument. If not, social considerations cannot plug the gap which rationality principles allegedly leave. We shall see that this suggests that the key issues underlying these matters are not those of rationality versus the pursuit of social interest but simply over the particular form which rationality principles take.

There is a widespread conviction that the procedures of logic and rationality guarantee no unequivocal answers to inductive and explanatory enquiries. One of the sources of this conviction is the so called "Quine-Duhem thesis", which is generally understood to entail that an agent faced with a seeming difficulty or contradiction between data and theory has a free choice as to whether he deems the theory falsified or retains it by parking the blame elsewhere within the system of knowledge with which the theory is connected. The significance of this claim is badly exaggerated by those who have an interest in that latitude of choice.[19] In fact the result is based on the deductive compatibility of any typical data and an isolated theoretical component of a theoretical system which comprises the tested theory, auxiliary hypotheses, initial and boundary conditions. But the fact that the data and deductive logic fail to constrain the agent to a unique decision says nothing about whether inductive logics or decision theories might not present unique prescriptions. Thus it might be argued that an attempt to preserve a tested theory in the face of a putative falsification by scrapping one of the auxiliary assumptions might, in inductive terms, be much more expensive than scrapping the theory itself. For example, if defending Newtonian theory from the difficulties presented by the anomalous advance of the perihelion of Mercury invited altering an estimate of the mass of the Sun, that choice would soon reveal immense costs in the damage to already successful applications to the major planets, comets, and perhaps all Solar System astronomy since, among other reasons, our understanding of the motion of the Earth, our observatory, would be called into question. Certainly deductive logic alone permits many freedoms, but inductive rationality would sound the alarm. In the light of this, it seems to me that once again the key issue is the existence and character of the particular inductive procedures which we regard as normatively compelling, and not the Quine-Duhem thesis. The Quine-Duhem Thesis only does the job for the new wave if certain claims about the underdetermination of inducitve, as well as deductive procedures are assumed. *That some inductive procedures are unequivocal is, however, an essential premise of the social interests explanation picture.*

Barry Barnes has claimed, in presenting the social intersts model of science, that in order to understand choices in science we "must turn to the goals and interests which inform judgment when concepts are applied and usage developed. It is by reference to goals and interests that particular modes of concept application, selected from and preferred to innumerable alternative options, can be made intelligible".[20] Although Barnes allows that "we lack a precise and detailed account of the relationship between goals and interests on the one hand, and concepts and beliefs on the other",[21] he is confident that when "a concept is applied, a particular is attached to one cluster of instances in preference to others. This preference must be taken to reflect a judgment as to which strategy of concept application best furthers specific goals, objectives, or interests".[22] And, of course, not only does this function at the level of individual epistemic acts, but also collectively and socially: "Similarly, general acceptance of the preferred strategy reflects the shared judgment that communal goals, objectives and interests are best furthered thereby".[23]

The social dimension to this choice model of epistemology is not, Barnes tells us, eliminable. It has an essential role, because, among other things, there is underdetermination in inductive logic. If we try to explain why a scientist chooses one generalization rather than another, we face the social element; "remove it and refer only to innate inductive propensities", he tells us, "and immediately acts of concept application and associated inferences are insufficiently determined. No account of inductive logic, or of an individual agent as an inductive learning machine, suffices to identify the 'best' way of applying a specific concept."[24]

Barnes has clearly identified the social choices inextricably bound up with cognition as, effectively, decision making under uncertainty. Deliberating agents face problems of optimization. They must choose which of a set of available courses of action will optimize their social interests. It is obvious however that if this is indeed the correct picture, the decision procedures which are followed by the agent must be exempt from the arguments for underdetermination which have been assumed binding on inductive logic generally; otherwise the interests answer is no answer at all to the question of scientific choice and it fails to plug the "underdetermination gap".

It is an extraordinary arbitrariness on the part of the new wave to imagine that, to put it bluntly, philosophers' inductive and decisive procedures are bound to be equivocal, while their own inductive decision procedures (avowedly unspecified) are not. For example, a case can be made that the social interest optimization account must be at least liable to the kind of criticism levelled at Bayesian probability theorists, that there is no rational way to provide *unique* values for the representation of uncertainties in decision making under uncertainty. If the new wave's (up until now merely hoped for)[25] models of incorporating judgments of uncertainty into decision making *can* yield unique values for degrees of belief, then they will have refuted themselves in their critique of inductive procedures. If they *cannot* then they have also refuted themselves, either by failing to give means by which they can represent agents making unequivocal choices in social interest optimization problems, or by spiking their enterprise on a regress. An Edinburgh savant decides to be a phrenologist, we shall say, because he computes that the course of action optimizes his expected social interests. We have no explanation of his choice if we cannot indicate how he arrives at a sufficiently unequivocal representation of the uncertainties concerning the consequences of his espousal of phrenology in his social milieu. If we say that the uncertainties are not represented uniquely, and once more say that the equivocation gap is plugged by a higher level social interests account, that account will in turn have its own uncertainties, and the regress continues. If we stop the account at any level, we concede unique representations of uncertainty for the agent, and grant the inductive logicians what they had hoped possible for their theories from the beginning.

In an important sense, therefore, inductive logicians and new wave sociologists have a common interest in finding ways around the underdetermination of inductive inference and decision making. In this sense, the demands of their own explanatory models require concessions by the sociologists to some traditional philosophical aspirations for induction, both logical and psychological. In the spirit of fairness and reciprocation, I will now turn to look at further ways in which inductive logic makes certain concessions to contextual diversity, and suggest a way of looking at induction as "behaviour" rather than "inference" which should be congenial to the social interests approach in that interests are expressly identified in the inductive decision procedures.

Inductive Behaviour and Scientific Interests

A worthwhile project for a new wave sociologist, it seems to me, would be an examination of the prevalence of the methods of confidence and significance level tests in the social sciences. The case would be instructive because of the clarity with which one can separate out historical, causal explanations of why those tests, and in particular the use of 5%, 1%, and 0.5% significance levels for the rejection of null hypotheses, from the normative debate which rehearses the merits of using them compared with alternative approaches. It has been conjectured that the use of those threshold levels for hypothesis rejection is partly due to the widespread use of Fisher's statistical texts together with their statistical tables.[26] For convenience, simple round-figure significance levels were displayed, so that sample data could be interpreted in a direct, if approximate, fashion without having to bother with the computations of exact confidence levels which were so tedious in the days before calculators were widely available. Thus it was simpler to read off from Fisher's tables that one's sample fell into what is now known as the rejection region, i.e. was less probable than, say, 5% on the null hypothesis, than to compute the exact probability of the sample. What was a convenience in Fisher's approach became a prescriptive norm in many sciences, no doubt reinforced by the ease with which the 5% threshold could be assimilated to a rather traditional conception of inductive acceptance or rejection rules. For some reason, 5% became the magic number at which one rejected the null hypothesis and determined that one's sample was significant statistically.

Whatever the merits of the historical conjecture, and I suspect that there is much substance to it, the issue of these statistical thresholds is of considerable importance for understanding what induction is. For the ease with which significance tests could be assimilated depended, I think, in part on the prevalence of a certain view of induction as an "inference" which leads from data to the acceptance of a theoretical conclusion, mimicking as it were deductive inference. A number of methodologists think that this conception of induction is a mistaken one, and that deductive logic is not a good model for inductives practices, whether real or ideal. These methodologists, commonly Bayesians, see inductive practices in terms of decision theory rather than logic in the strict sense. Their judgments are supported in part by the reasons which underlie the historic fate of the "inferential" understanding of significance tests, which is to be grounded in statistical decision theory. For statistical decision theory solves the problem of how we should pick the probability at which we reject the null hypothesis. It does so by demanding that the deliberating agent specify what practical course of action he is contemplating, what consequences are envisaged from each action, and what values are attached to those consequences. Following a suitable procedure, for example, the maximization of expected utility, the agent is led to pick a course of action as best.

More recent decision theorists prefer to dispense with the rejection rule approach, an undesirable "middle man", and go directly to the evaluation of strategies like: "Buy the sweater", "Market the product", "Perform the clinical trial", or even "Publish the results" For once seen through the eyeglass of statistical decision theory the act of rejecting the null hypothesis is equivalent to the practical choice. Once this is clearly recognized, the pressure to represent the procedure as "inferential" diminishes. Hence the statistician's invoking of the notion of "inductive behaviour" rather than "inductive inference".[27]

Carnap, and methodologists influenced by him, have to a great extent embraced this behavioural approach to inductive methodology,[28] since they dismiss the inferential, acceptance rule conception. They preserve the notion of Bayesian inference in the general sense which is employed by the statisticians (which is to say, the notion of rules by which degrees of belief are transformed with the incorporation of new data); but they deny that scientists are or should be involved in any form of acceptance. Their contributions have greatly influenced the prospects for finding a solution, not to the traditional problem of justifying inductive inference, but rather, to a class of distinct problems concerning the rational change of partial belief.

What is conspicuous in the rejection of the "logical" model by these statisticians and philosophers is that their attention to practical decision making requires the express consideration of the values which are attached to the outcomes. The rationality of inductive behaviour demands almost precisely what is being pursued by Barnes, for example, in his interests model of scientific development: the consideration of values in scientific deliberation.

For the Bayesians identify two important parts of the structure of decision making, the uncertainty or probabilistic part and the preference part or utility theory. In fact, the concept of utility which is used in contemporary utility theory is conceptually tied to probabilistic notions. The procedures which are employed in determining preferences base the indexing of utility for an outcome on the probability of winning a standard prize in a lottery deemed by the agent in question as equally preferable to the outcome. Thus the notion of utility which is most commonly employed is known as "probabilistic utility".[29]

There is of course much debate over fundamentals, and there are different competing accounts of the foundations of the theory. At the same time, however, there is increasing use of the approach at the applied level in business, medical, and strategic enterprises.[30] This is, of course, no surprise because these are the typical domains in which we are likely to find explicit values. That is to say, these are applied or technological domains. We can see that applied decisions are just the sorts of things which would be illuminated by a theory of practical deliberation like decision theory.[31] It is, equally, no surprise that an interests model, which decision theory surely is, works or at least promises to work appropriately in the applied domain.

The issue which must interest the philosophers and

sociologists of science is whether what is called "pure science" can be assimilated to the same sort of model. For the expression "pure science" is one which is intended to reflect the freedom of the enterprise from technological or other applied values.

Part of the critique (and polemic) against rationality models which has been offered by the new wave sociologists, is the claim that philosophers have been too much embroiled with a picture of science given in "totally abstract contemplative terms".[32] Exactly what is intended by this is never given precise expression, but the diagnosis well suits the complaint which the Bayesian has against traditional models of scientific procedures. For within this approach the view of induction as a form of inference (which incidentally, and paradoxically seems to be embraced by the new wave)[33] is set aside, and the model of practical deliberation is substituted. So the question becomes: What sort of practical activity is it that pure scientists undertake?

The Bayesian answers vary, where there are answers. Carnap was somewhat non-committal, although he insisted that one could tolerate the notion of pure scientists using acceptance rules in contexts where no practical issues were attached. Whether there are such cases is however exactly the bone of contention here. Some philosophers feel that they can combine the virtues of the decision theoretic approach with the general orientation of the inference rule tradition. They think that the Bayesian utility structures can be used but with cognitive or epistemic utility functions which characterise scientific preferences over various states of relief from agnosticism. Thus the sort of acts which are contemplated are acceptances of hypotheses or answers to questions.[34] What sense can be made of the notion of acceptance here is a difficult matter, and in my opinion a decisive one. However, cognitive utility theory, despite its idealistic presuppositions, may be suggestive for elucidating the interests model in its pure form.

Barry Barnes' formulation of the interests model, contrary to widespread belief among philosophers, seems not to be necessarily committed to the solely extraneous social interest picture.[35] That is to say, Barnes appears to allow that the interests model can be satisfied by cases where the interests are the esoteric, technical, and "internal" ones related to problem solving, the growth of theory and such related cognitive matters. What seems important, indeed definitional, for the interests model is that deliberation, theory choice, concept application all reflect the explicit consideration of values. Barnes clearly takes the view that "a narrow basis of evaluation produced by social interests with a properly scientific context-independent basis involving only completely general interest in prediction and control" is "utopian".[36] But the fact remains that such a lucky eventuality would still count for him as an instance of the interest model. Since epistemic or cognitive utility theory is confessedly ideal or "utopian", we need not exaggerate the conceptual gulf which lies between the philosopher's normative model and the sociologist's descriptive one here.

Some Bayesians are unhappy with the cognitive utility approach because they pessimistically believe that when the bulk of the idealizations are fleshed out the notion of acceptance of a hypothesis will remain as unintelligible as they take it to be at present.[37] This leads some methodologists to suppose that the activity which is characteristic of scientistis as such is the assignment to hypotheses of judgmental probabilities which can then be employed by others who have an essential interest in application of the knowledge to add their own problems with the relevant states, outcomes, acts and utility functions. This conception restricts the Bayesian scientists simply to Bayesian *inference*. Others suspect that there is more to scientific rationality than probabilistic inference and concur, at least in *their* sense of the terms, that scientists' practice reflects the pursuit of interests.

One such approach emphasizes the practicality of the acts which have hitherto been associated with the inference rule approach to induction.[38] The perspective taken here is that the traditional problem of induction conflated at least two different levels of analysis in what I shall call the *fallacy of pragmatic descent*. This can be seen in the context of Barnes' and Bloor's critique of "evidencing reasons". For the historical inquiry into the social and other proximate causes of scientific belief, essentially asks for answers at a pragmatic level. That is to say, they want answers couched in terms which make essential reference to the deliberating agent. (They also ask for descriptive rather than prescriptive answers, but that is to cut the pie in a different direction from the one entertained here.) The philosopher is typically concerned to answer questions about evidencing reasons in terms which are essentially semantic. They concern the relations between statements or propositions, and what relations of implication hold between "premises" and "conclusion". Strictly speaking such an enquiry will not even refer to the historical individuals whose deliberations may have sparked it.

Now, one common convenience in the representation of logical relations, which may generally be labelled semantic, is their representation in pragmatic terms. Thus questions about implications might yield answers given in terms of *inferences* strictly understood. For example, although a semantic answer might say that p implied q, a pragmatic expression of it might be given, such as "q may be inferred from p". The latter, of course, implicitly makes reference to an individual (maybe all individuals) since, unless we are outflanked by neologism, it is agents who infer, while propositions imply. For the most part, the permeability of the boundary between implication, the semantic notion, and inference, the pragmatic notion, is innocuous, and frequently, as I have indicated, a convenience. But it does in certain respects suggest the conflation of methodologically important kinds which must be firmly distinguished.[39]

When we address the problem of inductive implication (which by analogy with the deductive case we may call it "the semantic problem"), we attempt to formulate and rationalize true relations which hold between propositions. In the deductive case, these generally take the form of statements such as:

If p is true, then q is true,

or

If p, then q.

These provide the basis for so called rules of inference. In fact, not infrequently they themselves are identified as the rules of inference, but they are not the rules of inference strictly speaking. The latter have the character:

> If you know p, and know p implies q, then you are warranted in accepting or believing q.

It is plain that the former "rules" are semantic truths concerning abstract linguistic objects, while the latter is a rule of rationality which applies to the epistemic "acts" of agents. Traditional concern among philosophers for the problems of logic therefore warrants the new wave's description as "abstract" but not "contemplative", since contemplative is a pragmatic notion. It also explains why the pursuit of evidencing reasons ignores agents and history; for the philosopher is after the logical or semantical relations which hold between abstract, timeless entities.

In the inductive domain, the issue of separating the semantic or implicational problems from the pragmatic or inferential ones has not been well attended to. It is therefore important to ask what form in particular the inferential problem takes. Commonly, inference is taken to involve acceptance or belief. But this has proved very hard to make sense of, as I noted in my remarks about the cognitive utility approach to decision theory. It does not mean "take to be true" in the sense of giving a subjective probability of unity. In any case, the assignment of a judgmental probability was one of the decision inputs, it is not one of the outputs (inference is not just probability kinematics). It cannot intelligibly mean "put your conclusion into your body of knowledge" *except* perhaps when that is understood at a level of practicality and concreteness which is almost unfamiliar in these matters. For what is being proposed here is that inference by an agent is the performance of a concrete act, sometimes private but mostly public, of the assertion of a conclusion.[40] The assertion may be oral, as in a speech act, or it may be written in, as Lakatos once put it, a "book act". Nor does it have to be simply the assertion of the judgmental probability of the propostion in question. That was an input to the decision to speak or write, but the output is a categorical assertion devoid of all explicit indicators of uncertainty. Inference on this account, in its strictest sense of what the agent does is a theory of the rationality of "speech" and "book acts"; and what is decided is publication in some form or other, even a clumsily described "refereed journal submission act".

This approach answers the question which faced the Bayesian decision theorist when he asked "what practical act is performed by the seemingly impractical, 'pure' scientists?"[41] It answers it in a way which is conformable to decision theorists' rejection of the deduction-mimicking "inference rule" or "acceptance rule" appraoch; and it does so in a way which is suggested by commonplace and traditional ways of distinguishing inference and implication, strictly construed. Further, it answers it in a way which should be in certain respects convergent with the aspirations of the new wave interests theorists. It leaves well enough alone in the semantic domain and yields to logicians the pursuit of truths concerning linguistic entities and their relations, while giving concrete interpretation to the decisions which must strictly be labelled as "inferential". If decision theory is to be seen as the most promising vehicle for rationalizing deliberations about those concrete acts, then the interests model has the day.

Whether the interests are ever or always social, in whatever sense is important to the new wave, is however an open question which has to be examined empirically. Some familiarity with the way that scientists vary their "speech acts" depending upon the context readily prompts the judgment that institutional structure, for example, enters in. A scientist who has some particular *unique* judgmental probability concerning a new theory will say slightly different things in his bedroom, in conversation with close, discrete colleagues, in 'not to be circulated' manuscripts, in circulated manuscripts, in journal submissions, and on the podium giving his discipline's annual Presidential address. They may range from "I *know* it's true", through "This is highly promising", to "Despite significant uncertainties this bears examination." What the interests are in those contexts which, among other things, differentiate the chosen speech acts despite the unique personal probability, is a matter for speculation and further enquiry.

There remains too the problem of whether the operative interests *should* be operative, given the cognitive goals of the enterprise. Whether the purely scientific problems which phrenological theory was capable of illuminating would best be addressed by a scientist whose choices were motivated by the desire to optimize the economic and political aspirations of the Edinburgh middle classes may be open to debate. I dare say that most philosophers, wisely or not, have an opinion on that.

Summary

In the above, I have attempted to isolate some of the issues which are at stake in the debate between new wave sociologists of science on the one hand and philosophers of science and epistemologists on the other. In particular, I have taken issue with the view that *relativism* in the sense which the new wave puts forward has any direct significance for what philosophers really fear. What the sociologists discuss is what I have termed *context dependent diversity of opinion*, which is permitted by traditional logic, and in various novel ways by modern philosophies of induction, whether they be understood normatively or as idealized psychological accounts of rationality. In illustrating how context dependency enters into the philosophers' rationality models, I hope to show that the sociologists' attack on normative questions is gratuitous 'overkill', partly because they can have what they want without such destructive argumentation as is common in the literature. This is amplified in the second section, which attempts to reconfirm the distinction of philosophers' and sociologists' interests. What count as good reasons, in the normative or prescriptive sense, is quite independent of the historical causes or reasons which influence scientific belief and action. Of particular significance is the sociologists' appeal to the alleged underdetermination of scientific choice. I argue that the (social) interest model requires models of rationality which permit agents unequivocal choices.

The sociologists' espousal of the thesis of underdetermination of inductive choice is, I argue, simply incoherent. For if socially motivated choices are to be unequivocal, there must be models of deliberation which escape the arguments (if there are trully compelling ones) for underdetermination in general. Once this is clearly recognized, it is apparent that the underlying issue is over the specific models of inductive practice which are to be imported into our account of deliberation. In conclusion,

I draw attention to the decision-theoretic models which have appeared in the philosophy of science in the last quarter century, and which promise to displace traditional "logical" accounts of induction. These have the interesting property that they explicitly introduce considerations of values, preferences, or utilities into the structure of deliberation, which I can only imagine presents the new wave sociologists with confirmation of an interests model of scientific thinking.

Department of Philosophy, and
Department of History of Science and Medicine
University of Western Ontario

Notes

1. The "New Wave" is part of a movement in the sociology of science which insists that the locus of social control in science operates on the esoteric scientific artefacts and activities of scientists, rather than simply at the larger institutional level. Thus the movement eschews the Mertonian research programme which, it is alleged, attends only to scientific values which are "content neutral" in that they are independent of the particular scientific artefacts and procedures with which the researching scientist is embroiled. With a little qualification, their views resemble Thomas Kuhn's thesis that the technical problem solving "exemplar" is the embodiment and implicit vehicle for scientific values and criteria of judgment and choice. In this paper I shall use the term more restrictively for the radical wing, in particular, Barry Barnes and David Bloor, and their cohorts, who have taken a particularly vehement anti-philosophical stand. The threat of the intrusion of sociology into the domain of cognitive rationality has been regarded by many methodologists with an anxiety out of proportion to its likely bad consequences. Also the likely good consequences have been almost entirely neglected. Part of the difficulty, I am convinced, is the set of far too extreme concessions demanded by Barnes and Bloor. The meat of their position does not require

this sort of expense, and indeed in some instances their own position would be undermined by their radical argumentation.

2. That is not to say that the issue is not discussed. It is rarely defined, however. See, for example, Augustine Brannigan, *The Social Basis of Scientific Discoveries*, C.U.P., Cambridge, 1981, p. 78. (Brannigan appears to consider relativism to be more a methodological (heuristic?) recommendation than an "ontological" thesis. Cf. David Bloor, *Knowledge and Social Imagery*, Routledge and Kegan Paul, London, 1976, p. 142, for a hint of "methodological" relativism.) An attempt at the explicit characterization of relativism by Barnes and Bloor is discussed below.
3. The following distinctions are modelled on William Frankena, *Ethics*, Prentice Hall, Englewood Cliffs, N.J., 1973, p. 109.
4. Barry Barnes and David Bloor, "Relativism, Rationalism, and the Sociology of Knowledge", in Hollis and Lukes, (eds.) *Rationality and Relativism*, Blackwell, Oxford, 1982, p. 23.
5. Larry Laudan, "The Pseudo-Science of Science?", Phil. Soc. Sci., 11 (1981), 173-198, (Reproduced in this volume.) See also his "Oversestimating Underdetermination: Caveats Concerning the Social Causes of Belief", ms., 1980. For my present purposes I shall waive Laudan's argument that logical relations are independent of causal efficacy, and will treat contemporary theories of inductive rationality as proto-psychological, i.e., causal mode of belief acquisition or "inductive propensities". Cf. Barry Barnes, *Scientific Knowledge and Sociological Theory*, Routledge and Kegan Paul, London, 1974, p. 161. n. 3.
6. For definiteness, I assume material implication.
7. A convenient elementary review may be found in Henry Kyburg, *Probability and Inductive Logic*, Macmillan, London, 1970, ch. 5. See also Roger Rosenkrantz, *Foundations and Applications of Inductive Probability*, Ridgeview, Atascadero, Cal., 1981, section 1.3, and Carnap's own *The Continuum of Inductive Methods*, Univ.

of Chicago Press, Chicago, 1952.

8. For example, Isaac Levi, *Gambling With Truth,* M.I.T. Press, Cambridge, 1973, p. 86.

9. This point is forcefully made by Mary Hesse in "Bayesian Methods and the Initial Probabilities of Theories", in Grover Maxwell and Robert Anderson, "Induction, Probability, and Confirmation", *Minnesota Studies,* VI, Univ. of Minnesota Press, Minneapolis, 1975, pp. 50-105. Some dent may have been made in the scope of this claim by Jon Dorling's results in his review in *British Journal for the Philosophy of Science,* 26 (1974), 61-71. See also Dorling and Eddington in that journal, 27 (1975), 160-161.

10. Levi, *op. cit.,* pp. 67-68, 168-171.

11. Rudolf Carnap, "A Basic System of Inductive Logic, Part 1", in R. Carnap and R.C. Jeffrey, *Studies in Inductive Logic and Probability,* vol. 1, Univ. of California Press, Berkeley, 1971, pp. 51-52.

12. Barnes and Bloor, *op. cit.,* pp. 27-28.

13. Barnes and Bloor, p. 28.

14. Barnes and Bloor, p. 29.

15. Barnes and Bloor, p. 28.

16. Bloor, *op. cit.,* p. 4.

17. Bloor, p. 5.

18. For an example see Barnes' *T.S. Kuhn and Social Science,* Columbia U.P., New York, 1982, p. 110.

19. Although Barnes claims that Duhemian holism - that testing presupposes a network of theoretical assumptions is "one of the best-grounded findings ever to emerge from the philosophy of science", he rapidly slides to the view that "the results of the experiment can always be blamed on the inadequacy of a *different* hypothesis in the relevant system"; Barnes, *T.S. Kuhn and Social Science,* pp. 73-74, 106.

20. *Ibid.,* p. 103.

21. *Ibid.*

22. *Ibid.*

23. *Ibid.*

24. *Ibid.,* p. 110.

25. There are no detailed models of the relation of social

interest maximization to epistemology and scientific choice, even at the level of idealization which is to be found in the inductive logic literature.

26. "The reason significance levels are used is that tables were published by Fisher, from which we can easily tell whether the levels are attained. We are then often too lazy to work out the exact tail-area probability,.... When computers become more readily available the precise values will be used more often. Kendall mentioned that Fisher produced the tables of significance levels to save space and to avoid copyright problems with Karl Pearson, whom he disliked. Fisher's first two books on statistics dominated the market for over ten years; so it became fasionable to use significance levels. Why the preference for 5%?; perhaps because this level is often obtained when the null hypothesis is approximately *true*. At least 5% of experiments lead automatically to a publication!"; I.J. Good, "Comments", in V.P. Godambe and D.A. Sprott, *Foundations of Statistical Inference,* Hold, Rinehart, and Winston, Toronto, 1971, p. 513.
27. "Inductive behaviour" is Neyman's term. See his "'Inductive Behaviour' as a Basic Concept of Philosophy of Science", *Review of the International Statistical Institute,* 25, (1957), 7-22. A valuable overview of the key issues between classical and decision-theoretic approaches to statistics is to be found in Vic Barnett, *Comparative Statistical Inference,* Wiley, London, 1973. See also R. Duncan Luce and Howard Raiffa, *Games and Decisions,* Wiley, New York, 1957, section 13.10.
28. See Rudolf Carnap, "Inductive Logic and Rational Decisions", in Carnap and Jeffrey, *op. cit.,* and R.C. Jeffrey, "Dracula Meets Wolfman; Acceptance versus Partial Belief", in Marshall Swain, (ed.), *Induction, Acceptance, and Rational Belief,* Reidel, Dordrecht, 1970, pp. 157-185. A clear contrast of the decision-theoretic appraoch with traditional "inferential" ones may be found in C.G. Hempel, "Turns in the Evolution of the Problem of Induction", *Synthese,* 46 (1981), 389-404. There is now a large literature on belief change as probability kinematics. The best starting point for this literature, as

far as almost all recent developments, is Roger Rosenkrantz' *Foundations and Applications of Inductive Probability,* cited above.

29. See Luce and Raiffa, *op. cit.,* chapter 2, and Appendix A.
30. A full treatment of the Bayesian decision theory in the context of public policy, medical, and business applications is given in Ralph L. Keeney and Howard Raiffa, *Decisions with Mulitple Objectives,* Wiley, New York, 1976. See also R.O. Schlaifer, *Analysis of Decisions Under Uncertainty,* McGraw-Hill, New York, 1969.
31. An informed repudiation of such an enterprise is Henry Kyburg's "Two World Views", *Nous,* 4 (1970), 337-348. I have given some hints about how such a practical decision model of pure science might be tied to sociological, exchange-theoretic approaches in "Scientific Rationality and Local Progress", *Nature and System,* 2 (1980), 219-230.
32. Barry A. Barnes and Donald Mackenzie, "On the Role of Interests in Scientific Change", in Roy Wallis, (ed.), *On the Margins of Science: The Social Construction of Scientific Knowledge, Sociological Review Monograph,* Univ. of Keele, 1979, 49-66, p. 52. For confirmation and partial justification for their view, see Kyburg, *loc. cit.*
33. See Barnes, *T.S. Kuhn and Social Science,* p. 110.
34. See Isaac Levi, *op. cit.,* chapter 5, and "Acceptance Revisited", in Radu Bogdan, (ed.), *Local Induction,* Reidel, Dordrecht, 1976, pp. 1-71. Some related issues are raised in Paul Teller, "Zealous Acceptance", in the symposium on "Acceptance as a Basis for Induction", in L. Jonathan Cohen and Mary Hesse, (eds.), *Applications of Inductive Logic,* Clarendon, Oxford, 1980, pp. 28-53.
35. Barnes and Mackenzie, *op. cit.,* p. 54.
36. *Ibid.*
37. See Jeffrey's complaint about acceptance in "Dracula Meets Wolfman...", p. 183. For his radical "oddsmaker view" of scientists, see "Valuation and Acceptance of Scientific Hypotheses", *Philosophy of Science,* 23 (1956), 237-246.
38. This view may conveniently be labelled "Radical

Behaviouralism". Cf., my "Scientific Rationality and Local Progress" (note 31).

39. Charles Morris distinguished three fundamental categories in the theory of signs: syntax (syntactics), semantics, and pragmatics, corresponding to relations among signs only, relations between signs and objects denoted by them, and relations between signs, objects, and sign users, respectively. See his "Foundations of the Theory of Signs", in O. Neurath, R. Carnap, and C. Morris, (eds.), *Foundations of the Unity of Science,* vol. 1, Univ. of Chicago Press, Chicago, 1971, pp. 91, 99, and 107. "Semantic ascent" is the term applied to the replacing of discourse which uses sentences or terms with discourse which mentions them, in a semantically "higher" language which contains such semantical predicates as "truth" and "denotes". See W.V.O. Quine, *Word and Object,* M.I.T. Press, Cambridge, 1960, pp. 270-276. My introduction of the term "pragmatic descent" relates to the common tendency (on all our parts) to represent issues which are strictly pragmatic in Morris' sense (in that they make essential reference to sign users or deliberating agents) by means of descriptions which confine themselves to terms, statements, and their semantical relations to the world. The central "pragmatic descent" is the structuring of the pragmatic problem of inductive inference (strictly speaking) as if it were merely a semantic problem of inductive implication and, conversely, the employment of strictly pragmatic terminology to express problems of inductive implication (if there is any such thing!!)

40. Ian Hacking has taken an interesting stand about this point in his reevaluation of Neyman-Pearson testing. In addition, he has, as far as I can see, taken a step which is highly congenial from the point of view of this discussion. He has urged that we see the problem of induction as a theory of speech acts, in a way which reflects the crucial difference between the pragmatic and semantic problems in induction. See his "The Theory of Probable Inference: Neyman, Peirce, and Braithwaite", in D.H. Mellor, (ed.), *Science, Belief, and Behaviour,* C.U.P., Cambridge, 1980.

41. Hubert Blalock, in his *Social Statistics,* McGraw-Hill, New York, 1960, p. 135, pointed out that the only real decision facing a social scientist is whether to publish or suppress his findings. See James Skipper, et al., "The Sacredness of 0.05", in D. Morrison and R. Henkel, *The Significance Test Controversy,* Aldine, Chicago, 1970, pp. 155-160, p. 158.

David Bloor

THE SOCIOLOGY OF REASONS: OR WHY "EPISTEMIC FACTORS" ARE REALLY "SOCIAL FACTORS"

The critics of the sociology of knowledge have produced a variety of interesting responses and counter arguments. Their objections are launched from very different standpoints, and the profile of their concessions is different in each case. On the one hand there is the sympathetic treatment of Gutting, who accepts the 'symmetry' requirement for sociological explanation, but denies its relativist implications. On the other hand, there is the more sweeping dismissal of Jarvie, who wants to sever the mere beliefs of scientists from what he calls 'science as such'. Perhaps the line of critical commentary that contains most promise is that of Nicholas. His observation, that there is a considerable overlap between the 'interest model' of belief and recent work in decision and confirmation theory, is a valuable one. The programme of relating different inductive strategies to socially structured utilities is certainly one in which sociologists, historians and philosophers could cooperate.[1] Gaston's down-to-earth emphasis on the learning process indicates a similar point of contact between the disciplines. The most intriguing and colourful of the criticisms, however, comes from Butts. He argues his case with a detailed historical example -- that of Kant's account of unusual mental states such as extra-sensory percepetion. This not only provides a welcome relief from the abstract arguments that are in progress elsewhere; it is, as Barnes suggests, an admirable way to bring the discussion into focus.

Clearly this diversity presents me with a problem. A point-by-point reply is impossible. I have decided that my best course is to concentrate on *one* paper and deal with it as thoroughly as I can. This, at least, has the virtue that the confrontation involved is real rather than contrived. I have chosen for this pupose the paper by Ernan McMullin. (The only

J. R. Brown (ed.), Scientific Rationality: The Sociological Turn, 295–324.

respect in which I will allow my argument to break out of this framework is by the addition of an appendix in which I will say something about the case of Kant and E.S.P.) There are three reasons for concentrating on McMullin's paper. First, it formulates criticisms of a kind that are frequently aimed at the sociologist of knowledge. Thus, it overlaps with Lugg's conception of how intellectual and social considerations relate to one another. Second, despite the reservations voiced in the paper, there is much in it with which I can agree: for example, the discussion of implicit and imputed rationality. Finally, I have had occasion, elsewhere, to use and to admire McMullin's work, in particular his detailed analysis of Newton's ideas on force and matter.[2] So there is both common-ground and sympathy, and these increase the chance that an exchange of arguments will produce positive results.

McMullin's main line of argument concerns the reasons that are advanced for a belief. How is the sociologist to handle the inferences that his subjects make, or the move from the principles they espouse to the application of these principles? If McMullin is right, such matters frequently lie outside the scope of sociological analysis. McMullin calls experimental evidence, theoretical assumptions, logical appraisals and metaphysical principles, the "epistemic factors" in knowledge. His point is to distinguish them from "social factors" such as group membership and class interests. In what follows I shall continue to use his terminology. If we are to follow McMullin's meaning accurately, we must note at the outset that epistemic factors, unlike social factors, are to be thought of as verbal formulations. They have the status, he tells us, of propositions (p. 134). Accordingly, throughout his paper he equates them with 'principles', 'arguments', 'canons', premisses', and the like (pp. 129, 132, 136, 157). Although McMullin treats his epistemic and social factors as having different natures, and says that they call for different methods of investigation, he allows that they are frequently connected with one another. For example, the principles to which a group subscribes may be thought of as 'norms' and their association with the group may be explicable by 'tradition' and 'authority' and the processes of 'internalisation' (pp. 132, 136, 145, 157). These, however, are the maximum concessions that McMullin will grant the sociologist. Any further claims he sees as 'extreme' and 'question-begging' (pp. 147f, 154f). Thus he allows that when a social actor fills out the data of experience, and interprets them in a certain way, this *may* be socially explicable. But he will not allow that it *must* be open to social analysis. The interpretation may, he says, be guided by reasons, and these reasons may represent the ultimate level of explanation (pp. 144, 150, 152).

The focal point of McMullin's doubts concerns 'the interrelation of premises and conclusions' within the reasoning of the social actors under study. These, he says, 'must be explored to the fullest possible extent' (p. 134). But such explorations would not, in general, have a sociological character. The initial acceptance of a set of premises might be a matter for a sociologist, but not the subsequent move from premise to conclusion. This is, or can be 'purely epistemic' (p. 144). McMullin is careful to note that if the professed reasons why some conclusion is drawn are not the real reasons, then it may be that the real reasons are social, e.g. the reasoning may be under the control of class interest. With the exception of cases like this, there exists a realm of verbal behaviour where reason itself is operative. It is, we are told, an empirical matter for the historian to decide in any given case which of the two factors is responsible (p. 159). It is because he sees matters in this light that it appears to McMullin that the requirement of symmetry in the strong programme is a piece of arbitrary legislation. Does it not demand that we setttle *a priori* what can only be discovered by empirical study, by insisting that reasoning will always be governed by social factors *rather than* epistemic ones? (This, I stress, is how it looks to McMullin.) Within McMullin's two-factor model it makes sense to imagine circumstances in which the respective contributions of epistemic and social factors systematically alter. Control could pass progressively from, say, the social to the epistemic. This is what McMullin thinks has happened in the history of science. The 'insulation' of knowledge from the wider society has led to a progressive realisation within knowledge of 'autonomous' and 'transcultural' principles of rationality (pp. 153, 155-59).[3]

I wish to challenge this picture by arguing that epistemic factors are really social factors. The dichotomy that McMullin uses to give structure to his scheme seems to me a false one.

I agree that the reasoning processes of social actors, be they scientists, theologians or laymen, must be understood to the fullest possible degree. What I reject is McMullin's belief that such an understanding might be achieved without seeing that, and how, the link between premise and conclusion is socially constituted. What I mean is this: if we examine the propositions that make up McMullin's epistemic factors, we will not be able to understand the inferences that are made from them without appreciating their dependence on patterns of training, conventions of use, and the precise historical circumstances that attend their employment. These, I shall argue, are not merely conditions which attend or facilitate the operation of rational or logical principles. There is nothing else which, so to speak, lies behind our reasoning, to which our rational steps must correspond or have reference.

Consider, for example, what McMullin tells us about Descartes' principle that knowledge must be based on clear and distinct ideas (p. 130). This proposition states a metaphysical assumption and, because it was deemed important by Descartes, it is counted by McMullin as an epistemic factor which might explain some of his other beliefs. The proposition in question was the one that Descartes relied upon when he declared that the essence of matter lay in extension and that the idea of, say, attraction, was not admissable. This was what he offered as his reason. The question is whether we can really believe that the explanatory task is finished when we find that he rejected the idea of attraction on the grounds that it was not clear and distinct? Just look at the connection between reason and conclusion. How was the latter got out of the former? Adn if it is problematic why attraction is said *not* to be clear and distinct, it is equally difficult to see why extension *is* said to possess these virtues. According to McMullin, extension "was the clear and distinct idea, par excellence, since it defined the domain of geometry" (p. 130). So here is the reason for the attribution. But why should anyone attach any weight to this? It would not have been difficult to find a reason for saying that the domain of geometry was unclear, and hence for treating extension as obscure.

McMullin is right to say that if we are to understand Descartes we must grasp how important it was for him that scientific ideas should be made out to be clear and distinct. Certainly, we will not have understood him unless we know that he defended extension, and rejected attraction, by reference to this principle. My point is that a further problem awaits us: how and why was the principle applied in the way it was? I am not saying that perhaps Descartes was *wrong* to say that the idea of attraction lacked clarity. I do not think that we should address ourselves to such evaluative issues at all. We can get rid of evaluation by saying, for example, that it is *equally* right to apply the label 'clear and distinct' in the way that Descartes applied it, or in the *opposite* way. The problem is simply: why was it applied in *this* particular way?

The same observation could be made about McMullin's other examples. Consider, for instance, his discussion of the link that has been alleged between the Puritan movement and the growth of experimental science in 17th-century England (p. 134). What is important in this example for our purposes is not, of course, the acceptability of the original claim as it was set out by Merton, but the character of what, for McMullin, would be an acceptable explanation of the phenomenon if it existed. For McMullin the link, if it existed, would be satisfactorily explained if it could be shown that Puritans had reasons to be preoccupied with science in the light of their theological principles. We are told that we should not assume that Puritanism is a social factor of a non-cognitive kind, e.g. a diffuse 'value' or disposition. Rather, the Puritan involvement with science might be 'epistemically' explained by showing that it was implied by a set of theological premises.

Once again the problem is one of showing why the premises and principles in question were taken to have the meaning and implications that they were. Clearly, Puritan theology did not contain any principles of the form "Thou shalt experiment with air pumps". (Though even if it did, there would still be problems for McMullin.) In reality Puritan theology was cast in rather different terms. Such connections as were made between them and science had to be actively constructed by their users. Where 'social factors' come in is in explaining why particular implications were constructed in the way they were, and why different groups constructed different implications.

In order to show the importance of posing the question in

this way it may be useful to indicate, in general terms, how the study of the Puritan-science connection has developed since Merton's pioneering work. Setting aside detailed controversies over the definition and membership of the category of 'Puritan', the most significant fact to emerge has been the role played by the divergence between what may be called the radical and moderate Puritans. It transpires that the link between science and theological principle cannot be understood without putting it in the context of this dispute. The radicals construed their theological heritage in ways that emphasised the immediacy of their contact with God. The moderate wing presented God as more distant. Theologically, then, the movmement became divided and the point at issue was the question of the immanence or the transcendence of God.

If we take these theological principles out of context, and subject them to an abstract analysis, it might appear that the belief in God's immanence would be the one which would 'imply' the close, reverent study of nature. Surely, we could say, any move to put distance between God and his creation must correspondingly downgrade the material world and diminish the force of the injunction to lavish care and attention on its study. This line of reasoning would, however, produce an association between theological principles and scientific theory and practice which is the *opposite* of that uncovered by historical study. It was the moderate wing, around Robert Boyle, who advocated the experimental method and elaborated the corpuscular philosophy of nature which became central to the Newtonian world-view. For their part, the radicals, such as the levellers and the Diggers, had no particular concern with science, nor any commitment to experiment or to the corpsucular philosophy. Their knowledge was simple, practical and inspirational. So how was the actual connection forged?

The only answer that I know that has ever been advanced for these questions is one which treats 17th-century theological discourse as being, at one and the same time, discourse about God *and* the proper ordering of society. Roughly speaking, the doctrine of immanence provided the idiom in which the sectaries justified their radical political demands for autonomy, egalitarianism and individualism. If God is immediately known in revelation and by the inner light, then who needs priests, and by what right do others try to exercise authority over us? Those I have loosely called the 'moderates' became alarmed by the behaviour of the sectaries, and sought means to repudiate their policies. The previous alliances, derived from common interests, and sustained during the Civil War, began to break up. This generated intellectual and ideological activity. Rival policies had to be justified by rival cosmologies, designed to meet the very specific demands of the occasion. A greater stress on transcendence on the theological plane, for example, provided a means of justifying a more hierarchical structure than that desired by the sectaries.

Theories about nature and the proper basis of knowledge were part of this overall argument. They provided the middle term linking doctrines about God and society. In this way the political conclusions drawn from the theological premises could be further reinforced by being shown to be 'natural'. Take, for example, the issue of power and activity in nature and the question of whether matter was active or passive. If God was close and immediately present, then perhaps He could be identified with His creation. Nature itself was divine. hence it would be intrinsically active and self-organising, just as the sectaries wanted civil society to be self-organising. It was to oppose this picture, and its Levelling implications, that Robert Boyle and others insisted that matter was inert and inanimate and incapable of moving itself. The newly discovered dependence of matter on non-material active principles resonated with, and reinforced, the social patterns that were being advocated. Both nature and society were organised on an hierarchical basis with power at the top. An analogy was being created between the world natural and the world political. Similarly with the methods for gaining knowledge. The sectaries were 'enthusiasts'. Their critics sought to curb 'private judgement' by stressing the collective discipline of empirical enquiry and the experimental method.

When we realise that the theological principles, premises or canons that were overtly in dispute were whether, for example, God was the Soul of the World rather than the Lord of Creation, we can appreciate the complexity of the steps that had to be constructed to get from premise to conclusion.[4] We also have to appreciate the delicacy of the ideological balancing act that was needed by those who wanted to sustain the link between God and the new scientific approach. If they advocated too little transcendence they would be granting too much immanence, and hence identifying the Creator too closely with his creatures -- which is what the sectaries did. If they advocated too much transcendence, then it would hardly be an act of piety to study nature -- why not deduce its essential structure and principles from the Bible? Furthermore, the less the intermingling of God and nature, the more nature becomes a self-moving system, and the more does it again become like the cosmos used by the sectaries. There were, then, numerous difficulties surrounding the task of ensuring that the 'right' conclusions were drawn from the 'right' premises. The implications of any given position were frequently alleged to be different from those that their proponents intended.[5]

Even if some of the particulars of this story were challenged the point would remain that it is the right *kind* of account. It is addressed to the right problem. There is no way in which the historical links in question can be understood in terms of the unproblematic implications of a set of theological premises. The historically conditioned patterns of relevance assumed by the actors, and their innovative and selective employment of their cultural inheritance, both contrive to make that approach impossible. So, implications are themselves part of the historical problem. The unavoidable interpretive step is one that takes us beyond the body of propositions under consideration, out to the surrounding circumstances. The meaning, and hence the implications, of the epistemic factors in question can only be ascertained when we know the goals and interest that inform their use. We need to know what their users were trying to achieve with them, and the context in which they saw themselves purusing these goals. For this reason I suspect that McMullin is making a bad mistake when he says, of this very example, that whether a given set of beliefs "would count as reasons *outside* the group is of no concern" (p. 135). This is an invitation to ignore vital comparative data. If we want to understand a piece of reasoning then we must understand its differential credibility. Setting it within the framework of rival interpretations provided by other groups is a standard way of approaching the problem.

So looking at what goes on outside the group, as a way of illuminating what goes on inside it, *is* of vital concern.

Investigators who study the social background of epistemic factors -- at least, if they take their enterprise seriously -- are not pursuing the social *rather than* the epistemic. They are helping to show the social character *of* the epistemic. The sociologist of knowledge grants that there may be relations between, say, Puritan theology and science, and that these can be deemed to be logical implications. His aim is to show how these implications are created and sustained, or challenged and changed. In doing this, the verbal or conceptual links that can be exhibited cease to be explanations of the behaviour in question, e.g. performing experiments, or subscribing to the corpuscular philosophy, or joining the Royal Society. Instead one looks for the overall point of the behaviour, being careful not to throw away evidence by an arbitrary narrowing of the context. Historical sensitivity to the role of reasons thus involves something quite other than the decision, in each case, of whether epistemic or social factors have been at work. A deeper conception of the empirical task of historical enquiry can, in fact, discard this dichotomy. The aim becomes one of treating every inference as problematic, and every implication as something in need of explanation by the contingencies surrounding it. Can we really envisage this exercise being carried through in the case of *every* link between premise and conclusion? Is there a conventional, contingent, and in principle variable link, between every proposition in the set that makes up McMullin's epistemic factors? Perhaps social factors play a role when we draw out the consequences of vague metaphysical principles, or apply the tenets of some loosely formulated theological system, but can this be true of the more central steps in scientific reasoning? After all, McMullin distinguishes, albeit reluctantly, between standard and non-standard epistemic factors. Perhaps my claims only apply to the non-standard, or metaphysical side of our rational life. The standard requirements of logic and rationality, which have always been the main focus of philosophical concern, might represent an inner-core of epistemic factors that are safe from the sociologist. Thus, it may be said, there will be cases where it will be simply *illogical* to proceed in one way rather than another. Logical necessity demands that we draw certain conclusions, and here we are subject to a form of compulsion that has nothing to do with society. When describing a case of mathematical and logical reasoning, it would seem, all that the historian or sociologist can do is to check whether or not the actors are proceeding consistently. If they are, that is all that has to be said, except perhaps for explaining how they came to be in a position to reason correctly. If they are not being consistent, then their error or deviation may be illuminated by appeal to some antecedent social bias or limitation.

Whatever may or may not be conceded to the sociologist on the question of metaphysical reasoning, I take it that at the heart of the appeal to epistemic factors lies the issue of logical compulsion. Logical inference and consistency are epistemic factors *par excellence*. If this is so, then the whole question can be concentrated into one issue. Everything comes down to how we analyse the idea of following a rule. We can let the idea of a rule stand as a representative of all McMullin's epistemic factors. Explaining a piece of behaviour by saying that it is determined by a rule can then stand as a paradigm case of rational or epistemic explanation. In contrast to this, the strong programme in the sociology of knowledge is based on the idea that behaviour cannot be explained by reference to rules. It is, in fact, based on Wittgenstein's analysis of rule-following. This is how one formulation of the programme was first introduced in 1973 in a paper contrasting Mannheim's limited commitment to the sociology of knowledge with Wittgenstein's much bolder and more penetrating position.[6] I wonder if the critics of the sociology of knowledge realise that what they msut do to secure their position is to demonstrate that Wittgenstein's analysis of rule-following is wrong? So far as I am aware they have ignored this fundamental question.

Wittgenstein's point is that "no course of action could be determined by a rule, because every course of action can be made to accord with the rule" (PI, I, 201). This idea has been illustrated in an informal way, above, by saying that the idea of 'extension' could be classed as 'unclear' rather than 'clear', and by arguing that a positive attitude towards experimental science might have been associated with very different theological tenets from those actually espoused by men like Boyle and

Newton. But the point applies with equal force to *any* verbally stated premise -- and recall that epistemic factors are verbal statements. Any verbally formulated rule may be given what, judged by ordinary usage, would seem to be a deviant interpretation. If we try to overcome this permanent possibility by formulating another rule for the application of the first rule, then the words of our explanation might be given a deviant interpretation. For this reason, every rule and every explanation is, in the end, grounded in routine, habit and custom. As Wittgenstein put it: "'obeying a rule' is a practice" (PI, I, 202). Wittgenstein sometimes said of rules, just as he said of all meanings, that they are grounded in 'a form of life'. Stated less dramatically the point is that rules, like meanings, are *conventional*. And, of course, this too was a formulation that Wittgenstein frequently used. In saying this he did not just mean that *which* rule is adopted is a convention. Nor did he mean that the verbally formulated rule was itself the convention. What is conventional is the pattern of underlying judgments to the effect that certain acts or utterances are *instances* of a rule or are *in accord* with it. It is these socially structured judgements that constitute what Wittgenstein called the 'practice' of following a rule. A recent, and beautifully argued, statement of this position is Kripke's 'Wittgenstein on Rules and Private Language'.[7] Kripke uses a simple example from arithmetic and brings out, at least in general terms, the naturalistic and sociological implications of Wittgenstein's work on rule-following.

It may be objected that these considerations are highly abstract. Do they really have significance for the writing of history or the study of the sociology of knowledge? In fact the significance of Wittgenstein's arguments about rule-following is both pervasive and concrete. It shows itself whenever the users of a concept dispute -- or, for that matter, *don't* dispute -- judgements of "sameness". The conventional character of these judgements is no less significant if they bear on the concept 'polyhedron' or the rule for the use of the connective 'and', than if they apply to a concept like 'clear and distinct' or 'God'. This is a point that can be, and has been, illustrated in detail by examples drawn from the history of mathematics and contemporary discussions amongst logicians.[8]

Sociologists of knowledge who reject the distinction between epistemic and social factors wait with interest to see their critics' response to Wittgenstein and Kripke on rule following. This is where the argument ought to be taking place. If these issues were kept in the foreground it might lessen the frequency with which sociologists find their arguments mischaracterised by their opponents. As an example of a misunderstanding of this kind I shall now consider what McMullin says about the social explanation advanced by Steven Shapin for the debates over phrenology and brain anatomy that took place in Edinburgh in the early decades of the 19th-century.[9]

The first thing to notice is the manner in which McMullin formulates the issues that are to be discussed. There is in his paper no other conception of sociological explanation than one couched in terms of 'norms' and their 'internalisation'. For McMullin norms are principles which, if internalised, guide behaviour (p. 136). So once again, what is being assumed is some version of the rule-following model and the idea of explanation by rules. That there are alternative models, for example the interest model, and that Shapin may *not* subscribe to the idea of normative determinism, receives no recognition.

The second point is that McMullin proceeds as if he is dealing with a naive empiric who gathers data about the distribution of belief, and then brings his enquiry to a close, totally ignoring the reasons, justifications and arguments advanced by the actors. Obviously Shapin's account begins by locating the support and opposition to phrenology in different social groups within Edinburgh. But, says McMullin:

> When we have noted the correlation between support for phrenology and membership in the middle classes, we are immediately led to ask *why* the correlation should hold. Was there something in the *content* of phrenology that made it acceptable to middle class people? But this brings us right back to epistemic factors again. (p. 151)

This is offered by McMullin as a 'serious challenge' to Shapin and, indeed, to the strong programme in general. The impression created is that Shapin was *not* 'led to ask' why his empirical generalisation about the distribution of belief took the form it did, and that if he had, he would (or might) have been brought back to the content of the beliefs and *away* from social factors.

In reality the whole point of Shapin's argument is to explain the empirical connection. The general form of that explanation is that the connection arose because certain sections of the population seized on phrenology as a vehicle for their interests, an expression of their aspirations, and a justification of their discontents. Phrenology offered them a model of the mind, and a model of man in society, which gave them legitimacy and furnished them with an idiom in which they could understand themselves. It was because the doctrine could be formulated in a way that made it useful for these purposes, and because it *was* so used, that it was accorded the pattern of acceptance and rejection that Shapin uncovered. Using the language of 'homology', rather than McMullin's 'correlation', Shapin makes the point absolutely explicit in summarising his argument: "the homologies between cosmology and society derive from a context of use in persuasion and social management."[10] If the brain was like the phrenologists said it was, then that made it into a resource that could be polemically useful to the Edinburgh middle classes. The basis of the account is therefore the idea of *social use* -- a theme that receives no mention by McMullin, even though it explains the very thing that he said wanted explaining.[11]

What now of the rest of the objection? Is it correct to say that an explanation of the link between a social group and a belief "brings us right back to epistemic factors again"? Was there, asks McMullin, something about the content of phrenology that made it particularly acceptable to certain people? The answer is 'yes'.

It was, however, what that content could be made to signify in the Edinburgh context that was important. One feature of its content was that phrenology could be presented as a skill that could be rapidly acquired, easily validated and immediately applied.

However, it requires little ingenuity to see that it could be glossed very differently. For instance, it could take on the form of an esoteric art. As Shapin points out, the manner in which its claims were in fact treated ensured that phrenology carried implications about the range of persons who could acquire a knowledge of mental science. So interpreted, it could be used to justify a change in the boundaries of cultural participation. This meant a challenge to the status of the elite groups in Edinburgh who had previously been able to pronounce authoritatively on the nature of mind and its implications for morality and conduct.

Shapin's whole account, then, *hinges* on questions of content. It is crucial, though, to appreciate how he handles the content of phrenological doctrine. The propositions which make up the theory always have to be interpreted in their context of use. Content can never be dealt with abstractly because then, as I have indicated, the tenets of the theory become indeterminate, and could be interpreted in widely different ways. For example, as the doctrine was used in Edinburgh, the vague slogan that the brain is the organ of the mind was interpreted materialistically. It was used to oppose the separation of mind and body that was taken for granted by the university philosophers working in the tradition of Scottish Common-Sense philosophy. Similarly, both philosophers and phrenologists subscribed to the idea of mental 'faculties', but the philosophers used the idea abstractly and against a background assumption of the mind's basic unity. In the context of this dispute the phrenologists stressed the extent to which they went beyond generalised capacities like memory, and made reference to specific skills or attributes. They took the division of labour in the mind seriously, though in other contexts, for example in discussion with more radical, working-class phrenologists, they emphasised the hierarchical structure of the faculties. The republicanism of their mental model, like the republicanism of their politics, had its limits.

Perhaps the most important of all the observations to make on the content of the doctrine is the following: the Edinburgh phrenologists *changed* its content. The doctrine they received was not the one that they passed on. They altered it from a theory which emphasised innate, pre-determined personality characteristics to one which implicated the social environment in crucial ways. The environment was held responsible for whether an individual was able to do justice to his mental endowment. The innate *capacity* of the various mental faculties had to be understood in conjunction with the *energy* with which the faculty was exercised, and this was influenced by education and opportunity. Only when human potential was fully realised, said the phrenologists, would the workings of society be efficient and just. So the content of the doctrine was actively shaped and modified by its adherents. The meanings of the propositions in the doctrine did not determine the behaviour of its adherents. Rather, the aims and purposes of its adherents determined the meaning given to the doctrine. So there is no question of Shapin coming back to epistemic factors. Nothing about his procedure inclined him to leave them behind in the first place. They were central to his analysis all along, because they were the very things he was analysing.[12]

As well as the work of Shapin, McMullin also cites the work of another historian who addressed the Edinburgh phrenology debates. This historian, Cantor, was critical of many points in Shapin's analysis.[13] Indeed he proceeds in rather the way that McMullin advocates. What is it that divides him from Shapin? The main difference is that Cantor uses the philosophical and methodological preferences of the disputing parties to explain the debate, while Shapin asks how these differences might have arisen in the first place, and how they are sustained once they have appeared. Cantor explains the lack of agreement by reference to the incompatible and incommensurable ideas of the protagonists.[14] Shapin, on the other hand, explains the incompatibility of the ideas by the opposed interests of the parties concerned. Had these interests not been opposed, there would have been nothing to stop the tenets, principles and propositions of the rival accounts being rendered harmonious. Thus Shapin does not say that communication broke down because of incommensurable concepts; he says that the concepts were allowed to remain incommensurable because there was no concern with communication. This is not to ignore the role of ideas or the relations between ideas -- it is just to see them as part of the problem, rather than part of the solution.

The claim that this is the more correct stance does not derive, as McMullin suggests, from dogmatism. As I have tried to show, it rests on Wittgenstein's theory of meaning. It is but a practical application of the carefully argued theory that meaning is created by use, not use by meaning.[15] This theory of meaning also indicates the proper line of response to

McMullin's question of how we are to decide whether to appeal to "differences of social class or differences of belief" in explaining the course of, say, the phrenology debates. Even if social differences are basic, he asks, "how exactly would this be shown?" (p. 151). We have already seen that in practice the way to show the priority of the social can be very easy. We must notice how the group takes up a body of culture and changes it in a direction consonant with their perceived interests. This is what Shapin did and this is what other sociological accounts have done.[16]

One final point. Particular demonstrations and empirical studies are always vulnerable to the objection that they are special cases -- just as general arguments are always vulnerable to the objection that they are abstract or programmatic. This is why it is always important to work with both. McMullin formulates the 'special case' objection by saying that sociologists of knowledge are prone to support their case by a few oft-told tales and favourite examples. These examples, he says, represent a biassed sample "not at all adequate to the weight now being laid on it" (p. 154). The bias is said to lie in the fact that all the supporting cases "bear on the status of man and on the relations of man and society". McMullin instances studies of the Darwinian controversy and the Eugenics movement, adding that a "recent favourite is the Pasteur-Pouchet debate" (p. 154). There is some danger, then, that my reiteration of Shapin's argument will be wasted effort. It will be dismissed as a defence of just a single example within a narrow and unrepresentative class.

How far is the charge of bias justified? I think that it can be shown to be wholly unjustified. Indeed, McMullin's own references are sufficient to sow the seeds of doubt about his claim. Elsewhere in his paper he mentions the work of M. Jacob on Newtonian matter theory and J. Richard on the reception of non-Euclidean geometry in Britain. Although in neither case are the details of the analysis described, nor its sociological significance assessed, enough is said to reveal the fact that these examples are not subject to the limitations that McMullin attributes to the 'sample'. Nor does McMullin's description apply to other studies that can be found in *Natural Order*, the excellent collection of readings in the sociology of knowledge to which he makes reference. It does not, for example, fit Wynne's study of the Cambridge ether theoreists, or Dean's account of some long-standing debates in botanical classification.[17] Of course, a mere four exceptions to the rule would not suffice to make the sample respresentative. In reality, though, the inductive basis of the sociology of scientific knowledge goes far beyond any of the works listed above. A valuable resource for anyone wishing to appreciate the scope and variety of studies that shed light on the social processes involved in scientific knowledge is Shapin's 'History of Science and its Sociological Reconstructions'.[18] Perhaps a careful reading of the 149 articles and books in its (selected) bibliography would put an end to dismissive remarks about 'oft-told tales' and 'recent favourites'.

Appendix: The Case of Kant and E.S.P.

The material presented by Butts is so engaging that I cannot resist commenting on it. I may have some difficulty controlling my enthusiasm for speculation, but I will try to confine my remarks to the main point at issue. The question is whether, by the use of his historical example, Butts has demonstrated the priority and autonomy of what he calls 'the methods of intellectual history'. I think I can show that his analysis of Kant's reaction to Swedenborg does not work. It does not really explain Kant's response to E.S.P. and other unusual mental states. Indeed, I think that the example, as Butts presents it, helps to reveal the essential incompleteness of intellectual history, and its essential dependence at all points in information about the social context.

How are we to understand Kant's rather equivocal attack on Swedenborg's claims to inhabit not merely the everyday world, but also the spirit world? The methods of intellectual history require us, if we can, to see within the texts under study something like the formal structure of an argument. Where the inner structure of the reasoning is difficult to discern, as it is in this case, then the obscurity may be due to a suppressed premise. Before going outside the texts we should look for the clues which reveal "the enthymematic premise that functions as the meta-reason for Kant's troubled attitude towards Swedenborg and the paranormal" (p. 227). In this instance, the suppressed premise that is uncovered is Kant's deep commitment to the "double government methodology". This is a dualist doctrine stemming from Leibniz, which divides the world into a realm of necessity coexisting with a realm of freedom. The first is the world of mechanical and material processes governed by laws and forces; the second is the realm of spontaneous spirit and ethical significance. Knowledge of the first realm is through experience; knowledge of the second is, at least for Kant, of a distinctly sparse and intellectual kind. The double government methodology says that both realms are real and have their proper scope but are (more or less) self-contained. (Nevertheless, the world of spirit is held to be, ultimately, more important and more real.) The explanation that is then advanced to account for Kant's concern with Swedenborg is that the latter's contact with the spirit world, and his alleged feats of extra-sensory perception, would violate the double government methodology. Swedenborg's claims are said to treat spirits as if they could occupy regions of space and time, as if they could interact with material objects like human brains, and as if they had at their disposal forces, in the way that material substances do. All these assumptions would mix up the categories that the double government methodology is meant to hold apart. Hence "Kant had to be concerned with claims to E.S.P. For if there were spiritual forces, if spirit can be given habitation, then Kant's entire critical programme collapses in shambles" (p. 236). So Kant believed a certain principle; Swedenborg's claims conflicted with it, and therefore he produced his negative response.

As a sketch of how some parts of the episode stood relative to one another, this is perfectly acceptable. But as an explanation it has three disconcerting logical features. These call into doubt its claim to stand as a self-sufficient intellectual account, untouched by considerations of a social nature. First, any explanation by appeal to a methodological principle simply raises the question of where the principle comes from and why it has credibility. Second, when the principle in question represents a form of division of labour; when it is based on the metaphor of the relations between two governments, and when it clearly involves reference to questions of morality, the 'intrusion' of the social, as Butts calls it, is surely highly probable. Apparently, for Butts, if we can explain an action by reference to a reason, then the explanation is in no way sociological. This holds even if the reason that is advanced is a justification based on the need to protect public morality. I confess that I find the implied boundary between the intellectual and the social that is involved in this position somewhat obscure. Third, and most important, is there not an element of circularity in the explanation? We are asked to understand Kant's response to Swedenborg by appeal to the double governement methodology. At the same time, we can only understand what the double government methodology amounted to in Kant's hands by seeing how he dealt with Swedenborg. Remember that this methodology was continuously evolving throughout Kant's work. It was, as Butts says, accepted from Leibniz but *transformed* by Kant (p. 221), who was *searching* for the correct form of it (p. 226). So in the case of Kant's attack on Swedenborg we must understand the precise content of the methodology in terms of the action as much as we must understand the action in terms of the methodology. If this were not so, the proposed explanation would lose all specificity.

Let me develop this last point in a little more detail. If Swedenborg was attacked because his theory allows for interactions between spirit and matter, the habitation of spirit, and the existence of spirit forces analogous to material forces, then how does Butts account for the fact that the double government principle *already* allowed for these things? It is clear that Kant concedes these points elsewhere, and treats the concessions as consistent with the double government principle. So how can the principle as such explain the resistance to the further concessions required by Swedenborg's claims? The original concessions to which I am referring are, of course, those involved in fitting human beings into the double government principle. The principle represents a classificatory scheme with a boundary between matter and spirit. Every single human being conceived as a soul animating a body, then represents an anomaly. Each one of us is positioned precisely on the boundary line. Kant explicitly acknowledges the anomaly, but declares it an impenetrable mystery. E.S.P. and the mental states reported by Swedenborg in which he discoursed with spirits, merely represented a few more similar anomalies to the same classificatory scheme. But this time Kant adopted a quite different policy. He did not bow his head before their mystery. Instead he did his best to argue that seeing ghosts cannot be distinguished from delirium. Within the terms available to him from contemporary physiology he did his utmost to heap suspicion on these cases, and to indicate their dangerous and pathological character.

The difference between Kant and Swedenborg did not, therefore, turn upon the question of whether there was a boundary between two realms, those of matter and spirit. Nor did it turn on the question of whether that boundary could be crossed. The difference concerned the extent and number and location of the crossing points. It was a question of how intensely the boundary between the two realms was to be policed. Butts fails to put his finger on this point. His analysis in terms of methodological principles is too abstract. His procedure is like explaining why someone was stopped at the border between two nation states by saying that it *is*, after all, a border, and forgetting that other people are allowed to pass through.

I suspect that we will only understand the details of Kant's reaction to E.S.P when we realise that the themes being rehearsed are familiar ones to any anthropologist. Attitudes to trance states, spirit possession and other unusual mental happenings vary considerably from culture to culture. Sometimes they are viewed as benign, sometimes as dangerous; sometimes the access to these states is widespread, sometimes it is restricted to the occupants of special roles.[19] Control and access to spirit are variables, and the problem is to explain how any particular group comes to settle upon their preferred policy. Kant was advocating the strictest possible control. The terms on which we could have access were indeed common to everyone, but they were about as limited as they could be. We all occupied the spirit world, we were all part of a mystical body of spirits, he said, but we could have no experience of this fact during our life. Or, at least, hardly any. In C.D. Broad's account of *Träume eines Geistersehers* we are not only given a general account of the book's structure and argument, but some important passages are quoted at length.[20] In these passages Kant described what we can know of the spirit world. It transpires that although we cannot see our fellow spirits, nor talk to them, -- nevertheless some of our sense of belonging to that realm seeps into our life and behaviour. Do we not, said Kant, possess a mysterious inclination towards social and ethical behaviour -- a desire for community and cooperation with others? Here the laws of the spiritual realm are manifesting themselves in the realm of necessity.

> A secret power compels us to direct our intentions to the welfare of others or in accordance with the choice of others, although this often goes against the grain and strongly conflicts with selfish inclinations. From this arises moral motives ... the rigid law of obligation and the weaker one of benevolence, both of which extort many sacrifices from us. In consequence of this we perceive ourselves to be dependent in our innermost motives on the *rule of the universal will*. From this there arises in the world of all thinking beings a *moral unity* and a systematic constitution according to purely spiritual laws. (C.D. Broad, p. 138)

We can, he goes on, think of the phenomenon of moral motivation as "a consequence of a genuine active force by which spiritual beings influence one another" (Broad, p. 138). So the glimpse that Kant allows us to have of the spirit world we inhabit is one that reveals it to be the source of morality and ethics and benevolence structured by laws of rigid obligations. Where Swedenborg would claim to have easy, continuous, rich, varied and unproblematically sensuous access to the spirit world, Kant makes our moral inclinations the only form of access that we can routinely claim.

Why should Kant opt for a policy of high control and restricted access to the spirit world rather than adopting a relaxed attitude which allows spirit to wander freely abroad? The general form of the answer may be given by Durkheim's theory of religion. Durkheim said that when we worship God we are really worshipping an image of our social collectivity. God is society, and its basic principles of organisation define for us what is sacred. The peculiar attributes of spirits and divinities, he went on, derive from the fact that they represent in transfigured form the social forces that impinge upon us. They represent, for example, a reified understanding of our social obligations.[21] So we can see what is at stake when men dispute about the nature of spirits, and our access to them. An easy access to the realm of spirit means easy access to the means of justifying oneself. Restrict that access, and one restricts the scope of available justification. As you tighten control at the boundary around the realm of spirit you signalise the increased control of the individual. The dispute with Swedenborg must have carried for Kant a clear meaning about modes of acceptable community life. In defining the spirit world as he did, Kant was describing the social form that seemed to him most worthy of respect. No wonder that the principles of administration in the Kantian world of spirit had a certain Prussian austerity to them.

As it stands this theory has its disadvantages. It is easy to read in it the idea that Kant's position is just a passive 'reflection' of his society, or of the Pietist subculture in which he was educated. Fortunately, there is an easy way to refine Durkheim's theory and remove this defect. What we should look for are circumstances in which his stress on control has a clear point. Can we find an immediate *social use* for a doctrine which restricts access to the spirit world so sternly? Who was causing trouble? Were there, for instance, groups whose trances, ecstacies and claims to extra-sensory perception justified their lack of conformity to the rigid obligations of the moral law? There is, in Butts' paper, a hint that there was a target group that would lend more than a personal point to the attack on Swedenborg. In a footnote Butts mentions "the Schwärmerei, the irrational enthusiasm for things paranormal that seems to sweep through northern Europe in the second half of the century" (p. 244). Kant, we are told, saw this Schwärmerei as a source of corruption and inappropriate ethical and religious enthusiasm. The picture that is suggested is one of a popular movement associated with spiritualism that was seen as a threat to social order, and that might derive strength from Swedenborg's claims. In this context Kant's book would make sense. It would explain why, starting from premises that were remarkably similar to Swedenborg's, Kant set himself the task of sharpening up the differences in the precise ways that he did.

These hints are not developed by Butts. For him, they provide no more than local colour. Whilst telling us that no social factors intruded themselves he mentions Kant's concern for the corruption of public morals, and refers to waves of religious enthusiasm, -- but resolutely keeps these two claims apart from one another. And, indeed, in his scheme there is no connection. He has no theory that suggests how matters of morality and social order might bear upon the issue. But it is not part of my task to decide how much weight can be put on these facts. The question is not whether I can explain Kant's theory sociologically, but whether Butts can explain it non-sociologically. I am saying that Butts only *appears* to explain it, because he has an undemanding idea of what counts as an explanation. His proposal sheds no light on any of the nuances of Kant's relation to Swedenborg and, as I have tried to show, it is these nuances in the application of the double government methodology that constitute the real problem. By contrast the *kind* of account that I have sketched goes straight to these details. Whether an hypothesis of the kind that I have described can be sustained by further research I do not pretend to know. But what I hope is clear is that no explanaion which stops at a methodological principle can confidently lay claim to completeness and independence from social factors.

I have been cautious with my own speculations because of Butts' stern warning against sociological theories based on "unfounded generalisations" (p. 239). It is therefore worth drawing attention to the fact that the attempt to construct purely intellectual explanations itself encourages us to embrace unfounded generalisations. The essence of Butts' argument is that if Swedenborg was right, "the entire critical programme would be in shambles". If the emphasis is solely on the word "critical", this is, perhaps, true. By definition, the critical aspect of Kant's programme lay in its use as an instrument with which to restrict access to the spirit world. But would the *entire* programme have been in shambles? Surely not. The claim is falsified by the subsequent history of Kantian philosophy. One important element in that history was the shift towards a freer conception of the commerce between ourselves and the spirit world. The barrier between the phenomenal and the noumenal was dismantled, so that yet more carefully contrived insights could be vouchsafed us.[22] Think, for example, of the romantic movement. This represented in some respects, a clear shift towards Swedenborg's position. For Kant, Swedenborg's particular genius was thrown into doubt because it could not be distinguished from madness; for the romantics, madness was a sign of genius. The association was preserved, but the connotations reversed. So, once again, the relations between the two governments of Kant's methodology had been subject ot re-negotiation. Imagine reading the protocols of a real, diplomatic negotiation between two governments. Would we ever try to understand them without knowing what was at stake in terms of interest, power, prestige and control? *That* would be the method of intellectual history.

In presenting his example, then, Butts is dealing with topics that have long been of concern to the sociologist of knowledge. He is dealing with questions of purity and danger, boundary maintenance, category violation, response to anomaly, questions of differential access to spirits, problems of control and legitimation, and definitions of sickness. These are all social variables and involved the social use of nature. I would like to think that, after what I have said, he might be just a little tempted to leave behind references to methodology, and immerse himself in the rich literature on these topics. After all, there are hints in his paper that he is already in a position to bring his data into contact with facts drawn from the social context. He may, in fact, almost have within his grasp a sociological account of the beginnings of Kant's critical philosophy. How can he resist carrying his research forward?

Science Studies Unit
University of Edinburgh

Notes

1. The cross-disciplinary bridge is, in fact, already under construction. A major step in this direction has been taken by Mary Hesse with her 'network model'. This is essentially an account of the pattern of conventions that are found in systems of classification and concept application. Hesse then goes on to describe her network model in terms of probabilistic confirmation theory. See M. Hesse, *The Structure of Scientific Inference*, Macmillan, London, 1974. A point of particular interest in this account concerns the 'coherence conditions' of the network. These are not parts of the network but conditions external to it which result in certain parts of the verbal pattern of culture being held stable. In confirmation theory these can be modelled in terms of the *a priori* probabilities accorded to parts of the network; sociologically they can often be identified as social interests. For an examination of the network model from this point of view, see: D. Bloor, 'Durkheim and Mauss Revisited: Classification and the Sociology of Knowledge', *Studies in History and Philosophy of Science*, Vol. 13, No. 4, 267-297, 1982.
2. See 'Durkheim and Mauss Revisited', p. 290 for the reference to: E. McMullin, *Newton on Matter and Activity*, London, University of Notre Dame Press, 1978. The present exchange may be seen as a continuation of the difference of opinion that emerged in the above article. There I noted that McMullin sought to explain the Newtonian preference for the passivity of matter by appeal to metaphysical influence and theological tradition -- as if these things, in turn, were self-explanatory or self-perpetuating. I argued that he could, and should, have made reference to available historical work which explains these metaphysical and theological preferences in terms of their social use and the social interests that underlay that use.
3. McMullin suggests on p. 153 that I do not acknowledge or allow for the relative insulation of contemporary science from the wider social milieu. His evidence is

that I use Mary Douglas' idea that a scientist's perception of what is right and necessary in social relations generates his sense of what we will find necessary in nature. In fact, Douglas' principle in no way leads to a denial of processes such as professionalisation -- because these are themselves ways of structuring social relations. The process of professionalisation, and self-conscious 'insulation', led to some extremely interesting and characteristic accounts of nature. The literature on scientific naturalism is important here, e.g. F.M. Turner, 'The Victorian conflict between science and religion: a professional dimension', *Isis,* lxix (1978), 356-76. I discuss the topic in: D. Bloor, *Wittgenstein: A Social Theory of Knowledge,* London, Macmillan, 1983, Ch. 7, section 5, 'The Problem of Insulation'.

4. '[T]he word *God* is a relative Term, and has Reference to Servants, and *Deity* is the Dominion of God, not (such as a Soul has) over a Body of its own, which is the notions of those who make God the Soul of the World; but *(such as a Governor has)* over Servants'. Isaac Newton, General Scholium to the *Mathematical Principles.* This is William Whiston's translation given in F.E. Manuel, *The Religion of Isaac Newton,* Oxford, Clarendon Press, 1974, p. 16).
5. Later, when the High Church Hutchinsonians criticised the Newtonians, they did so on the grounds that the corpuscular philosophers collapsed God and Nature together. They saw the Newtonians as the Newtonians saw the sectaries, as making God the Soul of the World. C. Wilde, 'Hutchinsonianism, natural philosophy and religious controversy in eighteenth-century Britain', *History of Science,* xviii, 1980, 1-24.
6. D. Bloor, 'Wittgenstein and Mannheim on the Sociology of Mathematics', *Studies in History and Philosophy of Science,* Vol. 4, 1973, 173-191; reprinted in: H.M. Collins (ed.), *Sociology of Scientific Knowledge: a source book,* Bath, Bath University Press, 1982, 39-57.
7. S. Kripke, 'Wittgenstein on Rules and Private Languages', in I. Block (ed.), *Perspectives on the Philosophy of Wittgenstein,* Oxford, Blackwell, 1981, 238-312.
8. See the examples in: D. Bloor, *Wittgenstein: A Social*

Theory of Knowledge, London, Macmillan, 1983, Chapters 5 and 6.

9. S. Shapin, 'Phrenological Knowledge and the Social Structure of Early Nineteenth-Century Edinburgh', *Annals of Science,* 32, 1975, 219-243; S. Shapin, 'Homo Phrenologicus: Anthropological Perspectives on an Historical Problem', in B. Barnes and S. Shapin (eds.), *Natural Order: Historical Studies in Scientific Culture,* London, Sage, 1979, 41-71.

10. *Natural Order,* p. 63. On the notion of social use, see S. Shapin, 'Social uses of science', in G.S. Rousseau and R. Porter (eds.), *The Ferment of Knowledge: Studies in the Historiography of Eighteenth-Century Science,* Cambridge, C.U.P., 93-139.

11. On p. 42 of *Natural Order* Shapin had said: "The mere assertion that scientific knowledge 'has to do' with the social order or that it is 'not autonomous' is no longer interesting. We must now specify how, precisely, to treat scientific culture as a social product. We need to ascertain the exact nature of the links between accounts of natural reality and the social order". In other words, he had made exactly the point that McMullin was to make about the need to move from correlation to causation. I cannot imagine that McMullin would have developed his argument in the way he did if he had pondered on the significance of these passages.

12. To emphasize this point it is worth stressing that the most detailed account that we have of the technical content of the debate is to be found in a paper of Shapin's that McMullin does not cite: S. Shapin, 'The Politics of Observation: Cerebral Anatomy and Social Interests in the Edinburgh Phrenology Disputes', in R. Wallis (ed.), *On the Margins of Science: the Social Construction of Rejected Knowledge,* (Sociological Review Monographs), Keele University, 1979, 139-178; reprinted in: H.M. Collins (ed.), *Sociology of Scientific Knowledge: a source book,* Bath, Bath University Press, 1982, 103-150. In this paper the conflict of social interests is traced in the esoteric details of the subsequent disputes about the size of the frontal sinuses, the structure of the corpus

dentatum in the cerebellum, and the existence and precise location of the fibrous matter between the cerebellum and the brain stem. The refusal of either party in the debate to give in when confronted by the evidence adduced by its opponents widened the debate beyond simple matters of fact into esoteric questions about preferred anatomical technique and, ultimately, into proper methods of assessing evidence and testimony. The whole interest of the analysis is that it can and does illluminate these things. That is why the paper is properly called the politics OF observation, and not: politics OR observation.

13. G.N. Cantor, 'The Edinburgh Phrenology Debate: 1803-1828', *Annals of Science,* 32, 1975, 195-218; G.N. Cantor, 'A Critique of Shapin's Sociological Interpretation of the Edinburgh Phrenology Debate', *Annals of Science,* 32, 1975, 245-256.
14. E.g., "The most remarkable aspect of the Edinburgh phrenology debate was the high degree of incommensurability between the protagonists. They adhered to different theories and accepted very different classes of 'facts'. There was further incommensurability owing to major differences in philosophical, metaphysical and methodological principles, and also, in specific cases, in theological outlook. So many and so deeply held forms of incommensurability ensured that there was minimal interactions between the two parties". p. 217; cf. also p. 253.
15. Shapin's own appeal to Wittgenstein in connection with the idea of 'social use' is to be found in his 'Social uses of science' paper, p. 133. See footnote 10 above.
16. For example, Boyle changed the atomism of the Pagan thinkers of antiquity by removing from their atoms the power to move and organise themselves. Similarly, the ether theorists of late Victorian Cambridge changed the ether from a mechanical and material thing which might explain, say, electrical phenomena, into an electrical thing that might explain matter. In both cases there were technical *and* social goals that could be jointly furthered by these shifts. See for example: J. Jacob,

'Boyle's atomism and the Restoration assault on pagan naturalism', *Social Studies of Science,* Vol. viii, 1978, 211-233; and B. Wynne, 'Physics and Psychics: Science, Symbolic Action, and Social Control in Late Victorian England', in B. Barnes and S. Shapin (eds.), *Natural Order,* London, Sage, 1979, 167-186.

17. For Wynne, see footnote 16 above. J. Dean, 'Controversy over Classification: a case study from the history of botany', *Natural Order,* 211-230. Nor, in general, does the charge fit the large number of studies concerned with what has been called 'professional vested interests'. It is by studies of this kind that I would expect to find some sociological light shed on the esoteric debates within contemporary science, and hence on the example that McMullin cites of S matrix theory. For references to this theory see: A Pickering, 'Exemplars and Analogies: a comment on Crane's study of Kuhnian paradigms in high energy physics', *Social Studies of Science,* Vol. 10, 1980, 497-502, esp. p. 499; A. Pickering, 'Elementary Particles: Discovered or Constructed?', in W.P. Trower and G. Bellini (eds.), *Physics in Collision: High Energy ee/ep/pp Interactions,* Vol. 1, London, Plenum Press, 1982, 439-447, p. 445; and more generally, A. Pickering, 'The Role of Interests in High Energy Physics: the choice between Charm and Colour', in K. Knorr, R. Krohn and R. Whitley (eds.), *The Social Process of Scientific Investigation,* Sociology of the Sciences, Vol. IV, 1980, 107-138.
18. S. Shapin, 'History of Science and its Sociological Reconstructions', *History of Science,* Vol. xx, 1982, 157-211. Given the standard assumption that the sociology of knowledge must be (roughly speaking) entirely "externalist" in orientation, I would once again emphasize the importance of the section dealing with professional vested interests. This represents a form of "internalist" sociology.
19. M. Douglas, *Natural Symbols: Explorations in Cosmology,* Penguin Books, Harmondsworth, 1973, pp. 104-110; (e.g., "we tend to find trance-like states feared as dangerous where the social dimension is strongly controlled" p.

104).

20. C.D. Broad, 'Immanuel Kant and Psychical Research', in *Religion, Philosophy and Psychical Research,* London, Routledge and Kegan Paul, 1953, 116-155.
21. E. Durkheim, *The Elementary Forms of the Religious Life,* trans. J.W. Swan, New York, Collier Books, 1961. (First French edn. 1912.)
22. As representative of this trend, see: R.B. Haldane and J.S. Haldane, 'The Relation of Philosophy to Science', in A. Seth and R.B. Haldane (eds.), *Essays in Philosophical Criticism,* London, Longmans, Green and Co., 1883, 41-66, esp. pp. 51-2 and p. 62. This example, incidentally, shows that not even the entire *critical* thrust is lost.

INDEX

THE UNIVERSITY OF WESTERN ONTARIO
SERIES IN PHILOSOPHY OF SCIENCE

A Series of Books in Philosophy of Science, Methodology, Epistemology, Logic, History of Science, and Related Fields

Managing Editor:

ROBERT E. BUTTS

Editorial Board:

1. J. Leach, R. Butts, and G. Pearce (eds.), *Science, Decision and Value.* Proceedings of the Fifth University of Western Ontario Philosophy Colloquium, 1969. 1973, vii+213 pp.
2. C. A. Hooker (ed.), *Contemporary Research in the Foundations and Philosophy of Quantum Theory.* Proceedings of a Conference held at the University of Western Ontario, London, Canada, 1973, xx+385 pp.
3. J. Bub, *The Interpretation of Quantum Mechanics.* 1974, ix+155 pp.
4. D. Hockney, W. Harper, and B. Freed (eds.), *Contemporary Research in Philosophical Logic and Linguistic Semantics.* Proceedings of a Conference held at the University of Western Ontario, London, Canada. 1975, vii+332 pp.
5. C. A. Hooker (ed.), *The Logico-Algebraic Approach to Quantum Mechanics.* 1975, xv+607 pp.
6. W. L. Harper and C. A. Hooker (eds.), *Foundations of Probability Theory, Statistical Inference, and Statistical Theories of Science*, 3 Volumes. Vol. I: *Foundations and Philosophy of Epistemic Applications of Probability Theory.* 1976, xi+308 pp. Vol. II: *Foundations and Philosophy of Statistical Inference.* 1976, xi+455 pp. Vol. III: *Foundations and Philosophy of Statistical Theories in the Physical Sciences.* 1976, xii+241 pp.
7. C. A. Hooker (ed.), *Physical Theory as Logico-Operational Structure*, 1979, xvii+334 pp.

8. J. M. Nicholas (ed.), *Images, Perception, and Knowledge.* Papers deriving from and related to the Philosophy of Science Workshop at Ontario, Canada, May 1974. 1977, ix+309 pp.
9. R. E. Butts and J. Hintikka (eds.), *Logic, Foundations of Mathematics, and Computability Theory.* Part One of the Proceedings of the Fifth International Congress of Logic, Methodology and Philosophy of Science, London, Ontario, Canada, 1975. 1977, x+406 pp.
10. R. E. Butts and J. Hintikka (eds.), *Foundational Problems in the Special Sciences.* Part Two of the Proceedings of the Fifth International Congress of Logic, Methodology and Philosophy of Science, London, Ontario, Canada, 1975. 1977, x+427 pp.
11. R. E. Butts and J. Hintikka (eds.), *Basic Problems in Methodology and Linguistics.* Part Three of the Proceedings of the Fifth International Congress of Logic, Methodology and Philosophy of Science, London, Ontario, Canada, 1975. 1977, x+321 pp.
12. R. E. Butts and J. Hintikka (eds.), *Historical and Philosophical Dimensions of Logic, Methodology and Philosophy of Science.* Part Four of the Proceedings of the Fifth International Congress of Logic, Methodology and Philosophy of Science, London, Ontario, Canada, 1975. 1977, x+336 pp.
13. C. A. Hooker (ed.), *Foundations and Applications of Decision Theory,* 2 volumes. Vol. I: *Theoretical Foundations.* 1978, xxiii+442 pp. Vol. II: *Epistemic and Social Applications.* 1978, xxiii+206 pp.
14. R. E. Butts and J. C. Pitt (eds.), *New Perspectives on Galileo.* Papers deriving from and related to a workshop on Galileo held at Virginia Polutechnic Institute and State University, 1975. 1978, xvi+262 pp.
15. W. L. Harper, R. Stalnaker, and G. Pearce (eds.), *Ifs. Conditionals, Belief, Decision, Chance, and Time.* 1980, ix+345 pp.
16. J. C. Pitt (ed.), *Philosophy in Economics.* Papers deriving from and related to a workshop on Testability and Explanation in Economics held at Virginia Poly-Technic Institute and State University, 1979. 1981.
17. Michael Ruse, *Is Science Sexist?* 1981, xix+299 pp.
18. Nicholas Rescher, *Leibniz's Metaphysics of Nature.* 1981, xiv+126 pp.
19. Larry Laudan, *Science and Hypothesis.* 1981, x+258 pp.
20. William R. Shea, *Nature Mathematized.* Papers deriving from the third international Conference on the history and philosophy of science, Montreal, Canada, 1981. Vol. I, 1983.
21. Michael Ruse, *Nature Animated.* Papers deriving from the third international Conference on the history and philosophy of science, Montreal, Canada, 1981. Vol. II, 1983.
22. William R. Shea (ed.), *Otto Hahn and the Rise of Nuclear Physics,* 1983, x+252 pp.
23. H. F. Cohen, *Quantifying Music,* 1984, xvii+308 pp.
24. Robert E. Butts, *Kant and the Double Government Methodology,* 1984, xvi+339 pp.

SCIENTIFIC RATIONALITY: THE SOCIOLOGICAL TURN

THE UNIVERSITY OF WESTERN ONTARIO SERIES IN PHILOSOPHY OF SCIENCE

A SERIES OF BOOKS
IN PHILOSOPHY OF SCIENCE, METHODOLOGY,
EPISTEMOLOGY, LOGIC, HISTORY OF SCIENCE,
AND RELATED FIELDS

VOLUME 25